1986

X.25 EXPLAINED:
Protocols for Packet Switching Networks

ELLIS HORWOOD SERIES IN COMPUTERS AND THEIR APPLICATIONS

Series Editor: Brian Meek, Director of the Computer Unit, Queen Elizabeth College, University of London

X.25 EXPLAINED:

Protocols for Packet Switching Networks

R. J. DEASINGTON
IBM (UK) Laboratories Ltd.

ELLIS HORWOOD LIMITED
Publishers · Chichester

Halsted Press: a division of
JOHN WILEY & SONS
Chichester · New York · Ontario · Brisbane

First published in 1985 by

ELLIS HORWOOD LIMITED
Market Cross House, Cooper Street, Chichester, West Sussex, PO19 1EB, England

The publisher's colophon is reproduced from James Gillison's drawing of the ancient Market Cross, Chichester.

Distributors:

Australia, New Zealand, South-east Asia:
Jacaranda-Wiley Ltd., Jacaranda Press,
JOHN WILEY & SONS INC.,
G.P.O. Box 859, Brisbane, Queensland 40001, Australia

Canada:
JOHN WILEY & SONS CANADA LIMITED
22 Worcester Road, Rexdale, Ontario, Canada.

Europe, Africa:
JOHN WILEY & SONS LIMITED
Baffins Lane, Chichester, West Sussex, England.

North and South America and the rest of the world:
Halsted Press: a division of
JOHN WILEY & SONS
605 Third Avenue, New York, N.Y. 10016, U.S.A.

©1985 R.J. Deasington/Ellis Horwood Limited

British Library Cataloguing in Publication Data
Deasington, R.J.
X.25 explained: protocols for packet switching networks. —
(Ellis Horwood series in computers and their applications)
1. Computer networks   2. Data transmission systems
3. Packet switching (Data transmission)   I. Title
001.64'404   TK5105.5

Library of Congress Card No. 85-906

ISBN 0-85312-626-7 (Ellis Horwood Limited)
ISBN 0-470-20183-5 (Halsted Press)

Typeset by Ellis Horwood Limited.
Printed in Great Britain by R.J. Acford, Chichester.

Table of Contents

Chapter 4 — THE PACKET LEVEL

Chapter 5 — THE TRANSPORT LEVEL

Chapter 6 — 'TRIPLE X'

To my parents

Preface

When computer communications was in its infancy, in the mid- to late 1960s, the accepted strategy for the interconnection of computer systems was to create specialised modules to handle different functions. No overall scheme was used: each type of connection produced its own problems, for which specific solutions were developed. The areas addressed by these systems provided for the remote connection of terminals and remote job entry stations. This reflected the centralised view of data processing mainly held at that time. Computing resources were still very expensive and had to be shared among many, possibly geographically dispersed, user communities. The 1970s saw the creation by several major computer manufacturers of complete networking strategies designed to make use of common components in the systems design to allow the movement of computer communications systems away from the specialised to the generalised. Two typical systems of this type are the DECnet product from Digital Equipment Corporation, and Systems Network Architecture (SNA) from International Business Machines. These differ from previous communications systems in that they are both sets of standards for systems interconnection, rather than actual products themselves. They were, and are, used as corporate standards for the interconnection of systems and terminals for all the types of hardware and operating software produced by these companies. Thus if the purchaser of a networked system only used software and hardware supplied by a single manufacturer he could be sure to achieve the required connectivity. Owing to IBM's dominant market position the SNA protocols are widely emulated by other manufacturers to provide connectivity between their systems and those of

IBM. In 1977 the International Standards Organisation (ISO) set up a sub-committee to examine the development of a non-manufacturer-specific set of communications protocols to cover the range of requirements, from the link level protocols used on the communications line to detect and correct errors right up to the application level where the ISO determined to define a standard set of functions for file access, job transfer, terminal control etc. These could be implemented by all manufacturers to provide the buyer of a system supporting these protocols with the ready ability to set up hetero-geneous networks of computers. At the present time almost all manufacturers are committed to the development of software to implement the ISO protocols for Open Systems Interconnection (OSI) as each new layer is finalised. At the time of writing the lower four layers of the seven layer model for open systems interconnection developed by the ISO are completely defined, the upper three layers are generally very close to completion. The OSI protocols have recently become of some political importance since in the UK British Telecom (the privatised common carrier in the UK) and IBM attempted to set up a national Value Added Network making use of SNA protocols. The government has refused to licence this operation since it was considered that the public interest was not being served by the non-use of existent OSI protocols by such a powerful combination of companies.

The purpose of this book is to introduce the ISO seven layer model for Open Systems Interconnection in general, and specifically to cover the lower four layers which can be considered to provide the communications subsystem, the upper three layers being increasingly oriented towards applications. The first chapter gives details of the OSI layerings and the present state of their development; subsequent chapters proceed up through the bottom four layers until the Transport layer is reached (layer 4). At this layer we describe not only the ISO Transport Service but also some other Transport protocols at this level which are in widespread use. In the final chapter we cover another set of protocols which also reside at the fourth level, but which were defined prior to the main ISO work and are widely used for connecting terminals to host computer systems. This set of protocols, generally known as 'Triple X' has been included because of its popularity, even though it is not a true OSI protocol.

Finally I should like to thank my friends and past colleagues in the University of Strathclyde Computer Centre, and those in other establishments, for their help and encouragement in the development of the open system of communications in the UK academic community which was the inspiration for this book; and my wife Kate who has been the nidus of my motivation when it seemed that the standards were changing as fast as I could digest them.

R.J.D.

January 1985

Introduction to the ISO seven layer model

1.1 INTRODUCTION

The intention of this book is to introduce the reader to the protocols which form the bottom four layers of the ISO seven layer model. We also cover some other non-ISO protocols which are in wide use since they form *de facto* standards in themselves. In the past couple of years almost all the major computer manufacturers have adopted some form of layered architecture. Some attempt to describe their existing networking products in terms of the ISO layered model with varying degrees of success. The two major manufacturer specific architectures which predate the Open Systems Interconnection (OSI) model for data transmission are Digital Equipment Corporation and International Business Machines with their DECnet and SNA systems respectively. Both are now able to make some use of X.25 networks to form links in their proprietary networking systems, but not in a particularly well-integrated manner. The function provided by the X.25 network is ignored and the DECnet or SNA protocols are carried by the X.25 virtual circuits almost as if they were another physical medium with a different Link level protocol to provide the error-free characteristic required by the higher levels in the architecture. Many European manufacturers would like to break the IBM monopoly on communications protocols and are actively taking the OSI line towards a set of networking protocols which will allow the interconnection of any heterogeneous set of machines. In the United Kingdom the academic community have been largely responsible for the development of networking software for Open System

Interconnection by insisting that all computers purchased for academic computer services must have software and hardware for connection to a national Joint Academic Network (Janet). This network is based on the X.25 protocol together with several higher level protocols for Transport Service and several standard functions such as terminal access, file transfer, electronic mail and remote job submission. Since the higher levels of OSI are not yet standardised these protocols are temporary intermediate standards applicable only to the UK and a few other users. But they do however provide valuable experience in the practical implementation of such software which can be fed back into the standardisation process.

In common with the other architectures mentioned, the OSI protocols define what the interfaces are to be between various pieces of software or hardware, they do not specify how the internals of the software or hardware must be built. For example at the Physical level the V.24 interface specifies, directly and by reference to other standards documents, what voltages must appear on what pins of a particular layout of connector at what times and for how long in order to convey binary digits of data from a data terminal to a modem. It does not suggest how the voltages should be generated, or recommend a particular type of integrated circuit to be used. However, if the data terminal manufacturer and the modem manufacturers both correctly interpret the specification, then there should be no problems with the intercommunications of the two devices. This example, however, raises several problems. The fact that two disconnected groups of people will have have to work from a single document created by a committee made up of yet further unconnected people highlights the requirement for absolute accuracy and unambiguity. Whereas for an intra-company project the two groups could sort out between them what the specificiation ought to be, who is to be the arbitrator in this new situation? Even if two independently developed systems can interwork this does not in itself prove that they conform to the intention of the standard. What is required is the provision of an independent tester of protocols for conformity with the standard. Some work is being progressed on this in the UK but it is a non-trivial task.

The Department of Trade and Industry in the United Kingdom is sponsoring work to develop tests for the conformity of data communications protocols. However, the construction of a test environment by which the required sequences of messages can be generated is in practice often a greater problem. This is due to the very nature of the protocols in that they describe external manifestations of the protocols rather than the internal operation. Thus is it nearly impossible to create a transportable set of test routines since the interfaces to the different implementations of the same protocol layer are likely to be different in ways which depend on the nature of the operating system which is supporting the communications system.

1.2 A BRIEF INTRODUCTION TO OPEN SYSTEMS INTERCONNECTION

The International Standards Organisation (ISO) set up a new sub-committee in 1977 to develop standards for the interconnection of a heterogeneous set of computers. This is known as Sub-Committee 16, often referred to as SC16 in the literature. The objective of SC16 was the creation of a set of communications protocols which would allow different manufacturers' systems to interconnect and interwork provided they adhere to a set of standardised protocols required for 'Open Systems Interconnection'. The highest priority was given to the development of the overall architecture which would allow the construction of a set of layered protocols which could be easily expanded or altered to cope with future developments in technology in this area. The initial 'Reference Model of Open Systems Interconnection' was completed in 1979 and is widely known as the 'seven layer model'. Each layer is known as the Nth layer, the layers below and above being known as the $N-1$th and $N+1$th layers respectively. Each higher layer in the hierarchy is intended to provide value added over the more primitive operations provided by the layers beneath. Each layer will only communicate with the layers directly above and below it in the seven layer model, there will be no communications arbitrarily from one layer to another. The interfaces will be clean between each layer, the interface is defined by the protocol definitions of OSI, but not the internal functioning of the layer, except where it relates the input of one layer to its output at another. The interface is only detailed in so far as the data which must be passed is specified, not the exact mechanism for handling the interlayer interfaces. It may well be that in some implementations the operating system supporting the system will provide convenient message-passing mechanisms and interlayer communications may make use of this. Often, however, the operating system is less helpful and the inter-faces may be either of an *ad hoc* nature or non-existent with no boundaries between layers visible from outside the software imple-menting the protocols. The ISO determined a number of principles to arrive at a sensible layering of the functions which are needed by an all-encompassing set of communications protocols. In somewhat abbreviated terms they are as follows: there should not be so many layers that it is difficult to describe or integrate them; the layers should handle functions which are basically similar; the boundaries should be at points where experience has shown them to be correct, or where the descriptions of the services provided are small so that the amount of information to be passed is minimised, or where it may seem that the layer may provide some useful function in isolation in the future; the layering should allow for a change within a layer in future without having to disrupt other layers should a technology advance make such a change useful.

1.3 OVERVIEW OF THE SEVEN LAYERS

The Physical layer: This layer provides the electrical, mechanical and low level protocol for the establishment, maintenance and release of communications circuits to transmit binary digits of information from higher levels from one end of the link to another. Examples are V.24, V.35, and X.21. The level one definitions of OSI take these established physical layer standards and describe their exact use in an OSI environment.

The Link layer: Building on the communications function at the bit level provided by the Physical level this layer provides an error-free virtual channel over the error-prone layer beneath. The ISO has already defined standards in this area, HDLC in particular. The seven layer model expects that a subset of HDLC will be used as the Link layer for communications, but the possibility of using a character mode protocol is not excluded. As well as providing an error-free virtual channel by means of error detection and retransmission where errors have occurred, the link level also provides low level flow control so that the receiver is able to control the rate of sending of data, thus preventing flooding of the receiver.

The Network layer: The basis for all present work in this area is the CCITT defined X.25 packet-switching protocol. This is being enhanced to provide a more generalised addressing system which will allow for the possibility of interconnected networks in a 1984 revision. It is at this level that the basic multiplexing of data occurs. Many different virtual circuits may be operated over one link layer handling one physical connection between the network host and the local packet-switching exchange. The Network layer is also the level at which routing occurs between the communicating network hosts. Each intermediate node between the source and destination of a packet must decode the information up to the packet level in order to determine which node the message should be sent to in order to ultimately reach its destination. It is possible that in future other Network protocols will be developed to handle the types of networks which are termed 'connectionless'; these are typified by local area networks which generally operate on a 'datagram' basis. In these networks each message is a separate entity with no relationship to those which have preceded it or those which will follow it. This is opposed to the connection-oriented networks to which X.25 is addressed in which a virtual circuit operates between two communicating entities. Each message or packet which is inserted into the network is in a clearly defined order and so has some of the characteristics of a real circuit. The interface between the Network layer and the Transport layer above it is the point of convergence of these two networking architectures, the Transport layer assumes that it is operating over a connection-orientated network, though not necessarily over an error-free virtual circuit.

The Transport layer: This layer is used to provide a simple transparent connection between two higher level entities without them having to be bothered with the detail of which route to take, whether it is more cost-effective for this connection to share an existing Network level connection or to make a fresh connection. The Transport layer is equipped to provide various levels of 'quality of service' depending on the demands of the protocol layers above it. They may specify that a certain throughput is required, which might result in having to allocate more than one virtual circuit to carry the data at the Network layer and below, or that the required transit time of data passing through the connection is such that satellite connections should be avoided.

The Session layer: The main function of the Session layer is to provide services for activity management and interaction management. The former is concerned with the provision of synchronisation markers which can be used to indicate different phases of an application program or to allow them to recover, the latter is concerned with controlling who may send data at any given time. Each Session layer connection maps onto one Transport layer connection which is inherently full duplex in nature. Most applications, however, require that some discipline is imposed to control access to the link, a system of 'tokens' which give permission to send data is used for this.

The Presentation layer: This provides a common set of functions for the interpretation of data. Its functions may involve for example translating between the EBCDIC character sets used by some computer manufacturers and the IA5 character set defined by the ISO and used by many others. In a more general way it may be possible to exchange a complete definition of how the data sent and received is to be translated; a definition of a floating point number could be created so that the same number which might be represented in different ways on different systems could be sent between them. This activity is typically to be found in existing file transfer protocols which are able to deal with the varying character sets of different computer systems. This layer is not completely defined at this time but the outcome looks likely to be a compromise between the two extremes of simple translation tables and a complete data description syntax being exchanged by the two presentation services.

The Application layer: Protocols at this layer directly provide the user with the services they require. These can be divided into two classes, those which are user-provided application programs making use of the communications subsystem provided by the OSI systems beneath, and predefined applications which will be a standard part of the ISO system providing services such as file transfer, file manipulation, job transfer and manipulation, terminal access, electronic mail and so on.

1.4 THE CURRENT STATE OF OSI STANDARDISATION

The progress of a standard from its initial identification as a requirement through to the production of an international standard by the International Standards Organisation is a long and arduous path. In simplified form the sequence of events is as follows:

1. Unofficial work identifies the requirement for a new standard.
2. Technical Committee 97 approves a 'Work Item' and allocates it to a sub-committee.
3. Working groups produce repeated working drafts which are cycled until general agreement is obtained.
4. A working draft is accepted as a Draft Proposal (DP). A number is allocated to the standard at this stage.
5. After any changes made as a result of comments received the Draft Proposal becomes a Draft International Standard (DIS).
6. The Draft International Standard is voted upon by member countries and becomes an International Standard. This is usually a formality with no significant changes being made at this stage.
7. The International Standard may be subject to periodic revision.

The Reference Model for Open Systems Interconnection (Seven Layer Model) became a Draft International Standard in June 1982, and was promoted to International Standard (IS) at the end of 1983. Owing to pressure from the United States an addendum is being worked on to describe the use of OSI in 'connectionless' networks. These are networks that do not present to the users virtual circuits which simulate the properties of real circuits (data is received in the same order to that in which it was sent etc.) but use datagram networks, in which each message is treated as a separate entity without any particular relationship to other messages preceding or following.

The Network layer and below are largely formed by the X.25 definition produced by the CCITT, but are subject to some alterations at the third (network) layer since the existing X.25 definition was not completely satisfactory for direct use in the OSI environment, it had limited addressing space and could only send small amounts of Interrupt data (see Chapter 4). It was accepted as a DP in the second quarter of 1983 and is now a Draft International Standard. A Draft Proposal on the problems of worldwide addressing and the interconnection of such networks is expected at the end of 1984. The Transport layer has been one of the most stable layers and is now a Draft International Standard. The pressure to accept a 'connectionless' network service might require that the Transport Service would have to perform a bridging function between connection-oriented (virtual circuit) networks and connectionless (datagram) networks. This issue has not yet been resolved since it breaks the well-accepted principle that the network layer and below are responsible for the relaying function when data is being routed in an OSI network.

The Transport layer is now quite stable and there are in existence several implementations carried out by manufacturers of computer systems. The Draft Proposals 8072 and 8073 have been accepted without significant change and are Draft International Standards at the time of writing. As in the case of the network layer there is a problem caused by demands that connectionless networks such as some local area networks may have some impact on the Transport layer, as mentioned above, if the Transport layer is required to act as a bridge between the two fundamentally different network philosophies. The Draft Proposal for the Session layer was produced in early 1983 and is now a Draft International Standard. The Presentation layer became a Work Item in autumn of 1982, a Draft Proposal has been produced in 1984 but only for a small subset of possible transfer syntaxes.

The Application layer services became a Work Item in autumn 1983 but the architectural status of common Application layer services is still under consideration. The common protocols under consideration are:

— Virtual Terminal protocol, to provide a common view of all terminals connected to a network so relieving the application tasks of the need to handle each terminal type separately.
— Job Transfer and Manipulation protocol, to permit job desscriptions to be sent between machines on a network, allow the interrogation of thc various queues and be able to alter them remotely. It is questionable whether an extension of Remote Job Entry is really relevant in the increasingly interactive orientation of today's systems.
— File Transfer Access and Management protocol, to permit the exchange of files between machines in a network, the manipulation of directories and the remote access of records within a file.

The consideration of security is proceeding at all levels within the seven layer model, there are pros and cons for placing encryption in almost all the layers!

The Physical level

2.1 PHYSICAL LEVEL OVERVIEW

The Physical level in the seven layer model for open system inter-connection defines the interface presented by a network to a terminal permitting the exchange of data at the bit level. At the present time most data communications is carried on telecommunications circuits which were primarily designed for carrying telephone calls. In order to use them to carry data a modem is inserted at each end of the circuit which converts between the logic level signals produced at the interface of the computer or network and the analogue signals which are carried by the telephone networks. Two types of operation are available, the dial-up line, and the leased line. The signals presented to a computer connected by a modem to either of these types of telephone line are defined by a CCITT definition such as V.24. For packet-switched systems the leased type line is used, as the connection between the host computer and the network is fixed. In other types of operation the computer itself dials a number on the network which it knows to be connected to another computer, the remote computer senses by signals from the modem that an incoming call needs to be answered and can answer and communicate in a manner similar to the human usage of the telephone network. A development of this kind of operation is that in which a telecommunications authority builds a network particularly for computer communi-cations. If this is operated in a way similar to the telephone network then it too can present either the equivalent of a leased line or a dial-up network. The interface to be presented by fully digital computer communications networks is defined by CCITT definition

X.21. The definition of X.25 makes reference to X.21 bis, this is an interim specification to be used before the widespread introduction of X.21 networks and consists of a subset of the well-known V.24 specification. The definition of X.25 also defines the subset of facilities of X.21 that it uses. Since a connection to a packet-switched network is generally a fixed point-to-point link, an X.25 packet terminal will only use simple facilities. X.21 will become increasingly important as digital networks are installed to replace the old electro-mechanical systems in many countries, so we will describe the facilities of X.21 fairly fully here even though some of the more complex facilities are not applicable to the X.25 usage of X.21.

2.2 THE SIGNALS PROVIDED BY AN X.21 INTERFACE

Circuit Id.	Name	Direction DCE	DTE
G	Signal Ground	n/a	
Ga	DTE Common Return	←	
T	Transmit	←	
R	Receive	→	
C	Control	←	
I	Indication	→	
S	Signal Element Timing	→	
B	Byte Timing[a]	→	

a This signal is an optional additional facility.
DCE – Data Communication Equipment.
DTE – Data Terminal Equipment.

The function of the signals provided are now described before we proceed to a description of the protocol used between a terminal and a network providing the X.21 interface.

The signals provided by an X.21 interface are presented on a 15-way connector whose mechanical details and pin arrangements can be found in ISO document 4903. The electrical characteristics of the interface can be found in CCITT recommendations X.26 and X.27 which in turn refer to CCITT recommendations V.10 and V.11 respectively for signalling rates below and above 9600 bits per second. The description of the signals provided by X.21 is taken from the definition of these signals in CCITT recommendation X.24.

Signal Ground (G)

This conductor is used to provide a reference against which the logic states of the other circuits can be judged. It may be connected to the protective ground (earth) according to the dictates of the electrical installation regulations in force where the equipment is sited. If shielded cable is used to link the interface to the terminal the shielding may be connected to Signal Ground to reduce interference.

DTE Common Return (Ga)

This circuit is used in unbalanced-type (X.26) configurations to provide a reference ground level for the receivers within the DCE interface.

Transmit (T)

This circuit carries the binary signals which carry data from the DTE to the DCE. This can either be data during the data-transfer phase of a connection or call control information from the DTE to the DCE during the call set-up or clear-down.

Receive (R)

Binary signals from the DCE to the DTE are sent on this circuit which may be either data during the data-transfer phase of a connection or call control signals sent from the DCE during the call set-up or clear-down phases of a connection.

Control (C)

During the data-transfer phase of a connection this circuit will always be in the ON state. During call control phases this cicuit may be ON or OFF as defined by the protocol to be described later. This line is controlled by the DTE and indicates to the DCE the meaning of the data on the Transmit circuit.

Indication (I)

This circuit is used by the DCE to indicate to the DTE the type of the data on the Receive line. During the data-transfer phase the circuit is always in the ON state. During call control phases the signal may be either ON or OFF according to the protocol which we will describe later.

Signal Element Timing (S)

Signals on this circuit provide the DTE with timing information which permit it to sample the Receive line at the correct instant to determine whether a binary one or zero is being sent and also to present binary data to the DCE on the Transmit line at the correct instants so that the DCE can accurately recover the signals. This signal is provided at all times.

Byte Timing (B)

Signals on this circuit provide the DTE with eight-bit byte element timing. This circuit is normally ON but changes to the OFF state at the same time as the Signal Element Timing circuit signifies the last bit of an eight-bit byte. During the call control phases call control characters which are sent must be aligned with the Byte Timing signal. During the data-transfer phase the communicating computers may bilaterally agree to use the Byte Timing signal to define the end of each byte transmitted and received. The states of the I and C lines may be altered at any transition of the S circuit but the interfaces will only monitor and record changes in their status at the instant that the B circuit changes from the OFF to ON state.

2.3 X.21 LINE PROTOCOL OPERATION

Both the DCE and DTE may be in either a Ready or Not-Ready state. In the case of the DTE the Not-Ready state may be either Controlled or Uncontrolled. The Ready state for either the DCE or DTE is signified by the continuous transmission of a binary 1 on the T line for a DTE or the R line for a DCE at the same time the C or I line respectively is in the OFF state. The Ready state indicates that although there is not at present a call in progress a call could be accepted. The Not-Ready state indicates that the DCE or DTE is at present unable to accept a call. The DCE Not-Ready state is signified by the transmission of binary 0 on the R line while the I line is held in the OFF state. The DTE Uncontrolled Not-Ready state is indicated by signalling binary 0 on T line and holding the C line in the OFF state. This state signifies that the DTE is unable to accept calls owing to some abnormal condition such as a fault condition. In the DTE Controlled Not-Ready state the DTE signals a pattern of alternating binary 0 and binary 1 on the T line while maintaining the C line in the OFF state. This is used to indicate to the DCE that the DTE is operational but is temporarily unable to accept any incoming calls at this time.

We can summarise the above descriptions in a tabular form which we shall use throughout the detailed explanation of the X.21 protocol.

State	Circuit − R	T	C	I
DTE Ready		1	OFF	
DCE Ready	1			OFF
DTE Uncontrolled Not-Ready		0	OFF	
DTE Controlled Not-Ready		01...	OFF	
DCE Not-Ready	0			OFF

The subsequent description of the operation of this protocol is detailed from the point of view of the DTE end of the connection; thus an incoming call refers to a call which has originated in the network at some external location and is coming into the DTE from the DCE, similarly an outgoing call refers to a call made by the DTE to the network. As an aid to understanding the progress of calls we will take the analogy of the ordinary telephone network in our description of the various states which calls pass through.

In the definition of the protocol which follows the call control phases require that characters are sent between the DTE and DCE, the coding of these characters is defined by International Alphabet number 5 (IA5) as defined in CCITT recommendation V.3. All sequences of characters sent between the DCE and DTE must be preceded by at least two SYN characters in order to establish eight-bit byte synchronisation between the transmitter and receiver of the messages. In networks where the Byte Timing signal is provided the message preceded by its SYN characters must be aligned to the B line timing signals. Some networks will transmit SYN characters from DCE to the DTE and the DTE must align all call control messages with these SYN characters. Once a call is established only bit level synchronisation is enforced by the S timing signal, the two communicating DTEs may choose whether to give significance to the B timing signals or to remain in synchronisation with the SYN characters which were sent by the DCE in the call-establishment phase.

Call Request: The DTE will indicate its request to make a call by sending a continuous binary 0 on the T line and setting the state of the C line to ON. This is analogous to lifting the handset of a telephone prior to dialling a number.

Proceed to Select: The DCE indicates that it is prepared to receive calling information from the DTE. This is signalled by sending continuous '+' (plus) characters on the R line while the I circuit is in the OFF state. This compares to the normal response of the telephone system which provides a dialling tone in response to lifting the handset. This state is continued until the selection information is complete. As was mentioned above, and will from now be assumed, the signals sent on the T and R lines are always preceded by two or more SYN characters to establish synchonisation at the character level. The Proceed to Select signal will be sent within three seconds of the Call Request signal being made.

Selection Signal Sequence: The DTE transmits to the DCE with the C line ON. The Selection Signal must start within six seconds of the Proceed to Select signal being sent, and be completed within 36 seconds. The Selection Signal will consist of a Facility Request block or an Address block, or a Facility Registration/Cancellation block. The Selection Sequence is terminated by a '+' (plus) character. A Facility Request block consists of a code followed by a '/' (solidus) separator followed by a parameter value. Multiple Facility Request signals are separated by a ',' (comma). The Facility Request block is terminated by a '—' (minus). An Address block consists of one or more Address signals separated by ',' (comma). The Address signal may be either a full network address string or an abbreviated address in which case it will start with a '.' (point). A Facility Registration/ Cancellation block consists of one or more signals separated by '/' (solidus). These elements are Facility Code, Indicator, Parameter, and Address. Any trailing element which is not required may be omitted. Multiple Facility Registration/Cancellation signals may be sent by separating the signals by ',' (comma). The end of the Facility Registration/Cancellation block is indicated by '—' (minus) followed by '+' (plus).

A typical Selection Sequence might be as follows:

234241260106+   — an Address Block and terminator

The response of the network to a Selection Sequence sent by a DTE will be to continue to transmit '+' (plus) characters followed by Call Progress signals. The format of Call Progress signals is: a value, or values separated by ',' (comma), and terminated by a '+' (plus). The values will indicate that the call has been successful or that it has failed and the reason for the failure. The values of the Call Progress signals is shown in Table 2.1.

The Call Progress signal will be sent by the DCE to the DTE within 20 seconds of the end of the Selection Sequence. A typical

Table 2.1 – Values of Call Progress signals.

Code	Meaning	Category
00	Reserved	
01	Terminal called	
02	Redirected call	Successful
03	Connect when free	
20	No connection	
21	Number busy	
22	Selection Signal procedure error	Cleared owing to short-term conditions
23	Selection Signal transmission error	
41	Access barred	
42	Changed number	
43	Not obtainable	
44	Out of order	
45	Controlled not ready	
46	Uncontrolled not ready	Cleared owing to long-term conditions
47	DCE power off	
48	Invalid facility request	
49	Network fault in local connection	
51	Call information service	
52	Incompatible user class of service	
61	Network congestion	Cleared owing to short-term conditions
71	Long-term network congestion	Cleared owing to long-term conditions
72	RPOA out of order	
81	Registration/cancellation confirmed	Cleared by DTE procedure
82	Redirection activated	
83	Redirection deactivated	

Call Progress signal in response to a Selection Sequence requesting a call would be:

           01+   – signifying that the call succeeded

The Call Progress signals may be followed by DCE provided information. These blocks are also sent to the called DTE just after it accepts an incoming call. This will detail who is making the call and is terminated by a '+'. When the network has established a connection between the two DTEs it will signal to the calling DTE that the connection is Ready For Data by setting the I line to the ON state. There now exists a connection between the two DTEs until the call is cleared by one or the other.

We will now describe the appearance of an incoming call from the DCE to a DTE which is in a Ready state. The DTE will indicate to the DCE that it is in a Ready state by signalling continuous binary 1 on the T line and holding the C line in an OFF state. The DCE will indicate to the DTE that it is also in a Ready state by signalling continuous binary 1 on the R line an holding the I line in the OFF state. When a distant DTE makes a call on the DCE to call this DTE the DCE will signal to the DTE that someone is calling by sending continuous BEL characters to the DTE. This is analogous to the telephone system ringing the bell in a telephone set when someone is calling your number. The DTE Accepts the call by raising the C line to the ON state. This is equivalent to picking up the telephone handset in response to the bell ringing. The DCE will respond by sending DCE-provided information indicating who is calling. The DCE then indicates that the connection is established by setting the I line into the ON state. The call is then in the Ready For Data state. The Data Transfer state which is entered when both DTEs have entered the Ready For Data state may be terminated either by the DCE or by either DTE signalling a Clear. If either DTE wishes to initiate a call Clear it does this by signalling continuous binary 0 on the T line and setting the C line to the OFF state. The DCE will respond by signalling Clear Confirmation by sending continuous binary 0 on the R line and setting the I line OFF. This then is the equivalent of replacing the handset in an ordinary telephone system. The DCE will follow the Clear Confirmation by a DCE Ready signal, after at most two seconds, consisting of continuous binary 1 on the R line and the I line OFF. The DTE should respond to the DCE Ready signal by signalling the DTE Ready state of continuous binary 1 sent on the T line with the C line in the OFF state, this should be signalled within 100 milliseconds of the DCE entering the DCE Ready state. The DTE at the other end of a connection which was cleared by the DTE will receive a Cleared by DCE signal, as would both DTEs if the DCE cleared the call itself for some reason. The DCE indicates the Clear Indication to the DTE by sending continuous binary 0 on the R line and setting the I line to the OFF state. The

DTE will signal Clear Confirmation by sending continuous binary 0 on the T line and setting the C line OFF. The DCE will signal DCE Ready within two seconds of receiving the Clear Confirmation which the DTE should respond to by signalling the DTE Ready state within 100 milliseconds.

We have so far been describing the use of X.21 in its circuit-switched application. The same system will also allow for leased line operation. This is the mode of operation recommended by CCITT X.25 for the connection of packet terminals to packet-switching exchanges which we are primarily concerned with in this book. In leased line operation of X.21 the default state of the line is to be in the Data Transfer state. In this state the C and I lines are held in the ON state as in the circuit-switched situation. If a DTE wishes to terminate the Data Transfer state it does so by signalling continuous binary 1 on its T line with the C line in the OFF state. The DCE indicates termination of the Data Transfer state by signalling continuous binary 1 on the R line and setting the I line to OFF. The Data Transfer state is re-entered by the DTE setting the C line to the ON state and sending continuous binary 0 on the T line, the DCE signals that the Data Transfer has been re-entered by setting the I line to the ON state, at which point end-to-end data transmissions may recommence.

2.4 THE USE OF X.21 BIS, ALIAS V.24 OR RS232C

The leased line mode of operation we have just described is compatible with present V.24 operation if we substitute the terminology used to name the lines (see Table 2.2).

Table 2.2 — Correspondence between X.21 and V.24.

X.21 Name	V.24 Name	Description
G	101	Protective Ground
Ga	102	Reference Ground
T	103	Transmitted Data
R	104	Received Data
C	109	Control/Carrier Detect
I	105	Indication/Request To Send
S	114	Signal Element Timing
B	—	Byte Timing

The ability to interwork between V.24 DTEs and X.21 DTEs is defined by CCITT recommendation X.21 bis. In the leased line configuration the operation of the X.21 protocol is as described above except that at the DTE which is using the V.24 interface the 105 (Request To Send) and 109 (Carrier Detect) signals are substituted for and I and C lines respectively. For the circuit-switched mode of operation the mapping is more complex. As we said at the start of this chapter, the X.25 use of X.21 at the physical level is trivial, but the increasing importance of X.21 circuit switching in the future justifies its description here. As the transition to a full X.21 network will make many years, owing to the large investment in V.24 equipment, both by end users and by communications authorities, we shall describe here procedures recommended by X.21 bis for circuit switching also.

The protocol used in X.21 bis is essentially the same as that described by X.21 except that the V.24 interchange circuits are used instead of the circuits defined by the X.21 specification. We will therefore only describe the protocol in outline but with reference to the V.24 circuits used. Table 2.3 details the V.24 circuits that are used and the abbreviations we shall use to describe them.

Table 2.3 — V.24 circuits used in X.21 bis.

V.24 Name	Abbreviation	Description
102	G	Ground
103	TxD	Transmitted Data
104	RxD	Received Data
105	RTS	Request To Send
106	CTS	Clear To Send
107	DSR	Data Set Ready
108/1	CDSL	Connect DataSet to Line
or		
108/2	DTR	Data Terminal Ready
109	CD	Carrier Detect
114	TxClk	Transmit Clock
115	RxClk	Receive Clock
125	RI	Ring Indicator

Data which is used for signalling purposes, for example to send a network address for a call request, are transmitted as before preceded by at least two SYN characters to establish character synchronisation. The TxClk and RxClk signals are provided by the DCE at all times, both signals are derived from the same timing information in the DCE.

Table 2.4 describes the functions of these interchange circuits in the operation of the X.21 protocol.

Table 2.4 — Interpretations of V.24 signals under X.21 bis.

Circuit	State	Function in X.21
G	–	Reference and protective ground
TxD	–	Data transmitted to the DCE
RxD	–	Data received from the DCE
RTS	ON	DTE is in an operable state (Ready)
	OFF	DTE is inoperable (Controlled Not Ready)
CTS	ON	DCE is operable (Ready)
	OFF	DCE is inoperable (Not Ready)
DSR	ON	Ready for Data
	OFF	DCE Clear Indication or Confirmation
CDSL	ON	Call Request by DTE or Call Accepted by DTE
	OFF	DTE Clear Request or DTE Clear Confirmation
DTR	ON	Call Request by DTE or Call Accepted by DTE
	OFF	DTE Clear Request or DTE Clear Confirmation
CD	ON	DCE operable
	OFF	DCE faulty
TxClk	–	Clock supplied from DCE to control DTE transmitter
RxClk	–	Clock supplied from DCE to control DTE receiver
RI	OFF	Always except when an Incoming Call is signalled
	ON	Incoming Call from DCE

Call Progress signalling is not supported on X.21 bis connections. A provision is also made for manual answering of calls made from an X.21 DTE in that the call request time-out period can be extended from its normal two seconds to sixty seconds. It can be seen then that most communications interfaces capable of full control of a V.24 modem will operate with X.21 bis.

3

The Link level

3.1 FUNCTIONS OF THE LINK LEVEL

The Link level procedures specified by X.25 provide a mechanism by which data messages can be exchanged across a data link. Each message on the link is sent in a 'Frame', the Link level procedures define mechanisms by which those frames can be sent and received, with the detection of errors created by the Physical Link, their correction by retransmission, and the overall control of data being sent over the link. The HDLC subset used is called LAPB.

3.2 FRAME DELIMITERS AND TRANSPARENCY

All frames of information on the link are delimited by a 'Flag' sequence. This consists of a zero bit followed by six contiguous one bits followed by a zero bit. This sequence can never occur in the data of any frame because of the 'Bit-stuffing' transparency mechanism which is used. In this regime, whenever five contiguous one bits are sent, the transmitter inserts a zero bit, this is removed whenever five contiguous one bits are detected at the receiver, except for the case of a flag being detected. These functions are typically executed by the hardware of the transmitter–receiver. Whenever the link is idle, flags are continuously sent. The Physical level provides a bit-synchronous channel, the function of the flags is to provide Frame level synchonisation. If during the transmission of a frame the transmitter decides that the frame is not required or cannot be accepted by the other end, then it may abort the frame by the transmission of at least seven contiguous ones; again the hardware of the receiver would normally detect this and either completely ignore the frame if it is a D.M.A. device or inform the interrupt routine that the data received so far by a character-interrupt device should be discarded.

The last 16 bits before the final flag of any frame is defined as a Frame Check Sequence (FCS). This 16-bit sequence is defined to be the ones complement of the sum (modulo 2) of: The remainder of $x^k(x^{15} + x^{14} + x^{13} + \ldots + \ldots x^2 + x + 1)$ divided (modulo 2) by $x^{16} + x^{12} + x^5 + 1$, where k is the number of bits in the frame whose FCS is being generated, and the remainder after multiplication by x^{16} and division by $x^{16} + x^{12} + x^5 + 1$ of the contents of the frame between the last bit of the initial flag and the first bit of the FCS, but including neither. This function is generally implemented in the hardware of the transmitter and receiver. The transmitter is generally presented with a frame to transmit, the transmitter will add the leading flag, the FCS and the trailing flag as well as sending continuous flags when there is no data to be sent.

3.3 FRAME STRUCTURE

The first two octets of data after the leading flag are defined in HDLC to be Address and Control fields, after these an Information field may follow. The number of bits in the Information field is variable upto a limit agreed with the provider of the network service. The Information field will also contain an integral number of octets. The Address field will contain one of two values termed 'A' and 'B'. The value of 'A' is 3, (i.e. 00000011 in binary) and that of 'B' is 1 (i.e. 00000001 in binary). Value 'A' is used in the Address octet of frames containing commands from the DCE to the DTE and for responses to these commands from the DTE to the DCE. Address value 'B' is used when the frame contains a command from the DTE to the DCE and for the responses to these commands from the DCE to the DTE.

Thus the layout of a frame is as follows:

Flag : Address : Control : FCS : FCS : Flag

or, for frames with an Information field:

Flag : Address : Control : Information . . . : FCS : FCS : Flag

The Control octet defines the exact function of the frame, there are three basic types: Information, Supervisory, and Unnumbered. These are referred to as I-frames, S-frames, and U-frames respectively. The formats of these three types of frame are now defined:

Bits:	8	7	6	5	4	3	2	1
I-		N(R)		P		N(S)		0
S-		N(R)		PF	S		0	1
U-		M		PF	M		1	1

where   N(R) = Receive Sequence Count
        N(S) = Send Sequence Count
        S   = Supervisory function bits
        M   = Modifier function bits
        PF   = Poll/Final bit

The I-frame format is used to provide for information transfer across the link. The I-frame contains an information field.

The S-frame is used to perform supervisory functions on the link such as to acknowledge I-frames, or to request retransmission or suspension of sending.

The U-frame implements link control without sequence numbers, they are particularly used to bring a link up from a state where sequence numbers cannot be known, or error recovery.

The original use defined for the Poll/Final bit in the control field was to control the use of the physical line in half-duplex configurations. In these situations the master end of the line could send as many blocks as it wanted until a response was required from the other end, it would then set the Poll bit on to inform the other end that it now had control. When the other end has sent as much data as it had to send it would hand back control of the line by setting the Final bit. The protocols used for X.25 are always for full-duplex configurations so the Poll/Final bit is released for other functions. The Poll/Final bit is now used to assure the integrity of some responses, thus avoiding the confusion caused when an acknowledgement is lost through error; the response to a command with the Poll bit set must be a response with the Final bit set. The Poll bit is thus set on commands which require a response; a time is set within a valid reply must have been received or else the frame will be retransmitted.

3.4 SPECIFICATION OF CONTROL FIELD

The encodings for the HDLC control fields are now shown:

Format	Command	Bits:	8	7	6	5	4	3	2	1
I	I		N(R)			P	N(S)			0
S	RR		N(R)			PF	0	0	0	1
S	RNR		N(R)			PF	0	1	0	1
S	REJ		N(R)			PF	1	0	0	1
U	SABM		0	0	1	P	1	1	1	1
U	UA		0	1	1	F	0	0	1	1
U	FRMR		1	0	0	F	0	1	1	1
U	DISC		0	1	0	P	0	0	1	1

where the Format indicates either Information transfer, Supervisory, or Unnumbered functions.

The functions of the various types of commands are now given:

Information (I) is used to transfer data across the link at a rate determined by the receiver and with errors detected and corrected.

Receiver Ready (RR) indicates to the other end of a link that the receiver is ready to receive I-frames, it may acknowledge I-frames with an $N(S)$ sequence number up to $N(R)$ minus one. It is also used to clear a busy condition which was set by the transmission of an RNR.

Receiver Not Ready (RNR) is sent to inform the other end that no I-frames can at present by accepted owing to buffer shortage. The condition may be cleared by either UA, RR, REJ, or SABM. Normally a REJ is used as it specifies at which frame retransmission should recommence.

Reject (REJ) requests the retransmission of I-frames starting at the one indicated by the value in $N(R)$ which is to be the sequence number of the next packet required. It is sent when the receiver receives a frame whose $N(S)$ is out of sequence. It is also used to clear a busy (RNR) condition.

Set Asynchronous Balanced Mode (SABM) when preceded by a UA command in the link-down state indicates to the DCE that the DTE wishes to enter the link-up state. It is also used when in the link-up state to reset the link in both directions, i.e. the $V(R)$ and $V(S)$

state variables will be set to zero and any any outstanding frames will never be acknowledged.

Unnumbered Acknowledgement (UA) is used by either end to acknowledge the receipt of an Unnumbered command, i.e. either SABM or DISC.

Frame Reject (FRMR) is used by either end to indicate an error condition which cannot be recovered by the retransmission of identical frames. The frame has a three-octet information field following the control octet. The first octet contains a copy of the control field of the frame which is being rejected, the second the values of the $V(R)$ and $V(S)$ state variables of the transmitter. The third octet contains a code to indicate the type of error which caused the transmission of the FRMR. The arrangement of the information field of FRMR is now shown:

Bits:	8	7	6	5	4	3	2	1
	Control field of frame							
	$V(R)$			|CR|	$V(S)$			| 0
	0	0	0	0 |	Reason			

where CR is zero if the rejected frame was a command else one of it was a response, and Reason is defined as:

Bits:	4	3	2	1	
	0	0	0	1	— Invalid Control field
	0	0	1	1	— Invalid I-field
	0	1	0	0	— I-field too large
	1	0	0	0	— Invalid $N(R)$

Disconnect (DISC) indicates to the receiver that the sender is ceasing operation, the receiver will send a UA command and then enter the link-down state after interval T1. When the link is in the down state the DCE will send DISC at intervals of T1 to indicate to the DTE that it is up and ready to accept a link reset command.

The following table shows the responses that are required to each of the HDLC commands. A command with 'P' after it indicates the form of the command with the Poll bit set, a response with '(F)' after it indicates a response whith the Final bit set on.

Command	Response
I	I
	R R
	RNR
	REJ
I(P)	RR(F)
	RNR(F)
	REJ(F)
SABM	UA
DISC	UA
SABM(P)	UA(F)
DISC(P)	UA(F)
RR(P)	RR(F)
	RNR(F)
	REJ(F)
RNR(P)	RR(F)
	RNR(F)
	REJ(F)
REJ(P)	RR(F)
	RNR(F)
	REJ(F)

The functions of $N(R)$ and $N(S)$ within the control field are now described. These counters each occupy three bits and can thus store values between zero and seven. All arithmetic involved with these counters is thus modulo eight, that is, whenever one is added to a counter containing the value seven its value becomes zero. The software handling the link keeps four variables which control the flow of data. These are:

Receive Sequence Number N(R). This is the expected sequence number of the next received I-frame. Before the transmission of a frame containing $N(R)$ it is set to the value contained in $V(R)$ — the Receive State Variable. $N(R)$ indicates to the end of the link which receives it that all I-frames up to $N(R)$-1 have been correctly received.

Send Sequence Number N(S). This is the sequence number of an I-frame, before transmission the value is set to that of the Send State Variable.

Send State Variable V(S). This contains the value of the sequence number for the next I-frame to be transmitted. The value of V(S) is incremented by one after the transmission of each I-frame (using modulo eight arithmetic). Its value cannot exceed N(R) of the last received frame plus seven, which is the maximum number of outstanding frames permitted in this system.

Receive State Variable V(R). This contains the value of the next in sequence I-frame to be received. The value of V(S) is incremented by one (in modulo eight arithmetic) each time a valid, in-sequence I-frame with an N(S) equal to V(R) is received.

To understand the operation of the Link level a few further system parameters must be described:

Timer T1. The expiry of this timer will initiate the retransmission of a frame. Its value must be greater than the time taken to transmit a frame of the maximum size and to receive two similar frames plus the propagation time of the signals through the physical link. Typical values are between 100 ms and 10 s.

Timer T2. This is the maximum time which may elapse before an acknowledgement to a frame must be sent.

Timer T3. This is the period of time for which the DCE awaits a link set-up command before entering or re-entering the link-down state. Its value is set to the value of T1 times the maximum number of retransmissions N2.

Maximum Number of Bits in a Frame N1. This parameter is defined by the largest size of information field that will be carried by an I-frame plus the bits needed to carry the Address, Control, and FCS fields of the frame.

Maximum Number of Transmissions of a Frame N2. This defines the maximum number of times a frame will be transmitted, including its initial transmission following the expiry of timer T1, typically 20.

3.5 INFORMATION TRANSFER

The procedure for sending and receiving I-frames is now described, we shall assume here that the link is already in the up-state, the procedure to bring the link up will be described later. The remote end of a link indicates to the other end its willingness or not to receive data by the transmission of either Receiver Ready (RR) or Receiver Not Ready (RNR) frames. If the last received RR- or RNR-type frame was an RNR, then further transmissions should be suspended until an RR is received. When the transmitter has a frame to send (assuming that an RR is outstanding) it will send it with an N(S) equal to its current V(S), it then increments V(S), and sets N(R) to its current value of V(R). If timer T1 is not running when transmission commences it will start it. If the value of V(S) is equal to the value of the last N(R) received plus seven (the maximum number of outstanding frames) then further transmissions are suspended until a frame containing an advanced N(R) is received, excepting for possible retransmissions that are required.

When, and only when, an I-frame is received with correct FCS and a sequence number N(S) equal to the receiver's V(R), the next expected sequence number, the information contained in the I-frame is accepted and the data is passed to the Packet level handler. The value in the receiver's V(R) will be advanced to the value contained in the received packets N(S). If an I-frame is ready to be sent out by the receiver of the I-frame it will set the N(R) of the frame to be sent to the current value of V(R); if there is no I-frame ready to be sent then either a Receiver Ready (RR) frame will be sent, if the receiver has sufficient buffer space to handle new I-frames, or a Receiver Not Ready (RNR) will be sent; in either case the N(R) of the frame will be set to the current value of V(R). If an I-frame with an invalid FCS is received it is ignored. If the value of N(S) in the received frame does not equal the value of the receiver's V(R) the frame is discarded and a Reject (REJ) frame is sent with its N(R) one greater than the value of N(S) of the last I-frame that was recieved in sequence. Whenever an Information or Supervisory frame which contains an N(R) is received it is considered to be an acknowledgement of all frames up to the value of N(R) minus one. The timer T1 is restarted each time a value of N(R) is updated if there are still I-frames outstanding which require acknowledgement. If the timer T1 ever runs out the procedure for re-transmission is performed. When a REJ is received the transmitter sets its V(S) to the value contained in the REJ and commences to transmit the I-frame to which it refers. If other I-frames have been sent with an N(S) greater than the sequence number of the frame to which the REJ refers then these must also be re-sent. If the transmitter is already in the process of sending an I-frame it may abort the transmission by sending an abort sequence of seven contiguous ones on the physical link. If the transmission was sending either a Supervisory or Unnumbered frame when the REJ was received it will complete the transmission of the frame. If there was no transmission in progress

then the transmission of the rejected frames starts immediately. If a
Receiver Not Ready is received the timer T1 is started. If the timer
expires before a Receiver Ready is received a Supervisory command,
either RR, RNR, or REJ will be sent with the Poll bit set on. If no
response is received to this within a period of T1 the Supervisory
frame will be re-sent up to N2 times. After N2 retransmissions
without any response being received the link will be reset by sending
the Set Asynchronous Balanced Mode (SABM) command. This is
retransmitted at intervals of T1 up to N2 times, if no response is
received then the link enters the down state.

If the timer T1 expires at any time indicating that the remote
end of the link has not acknowledged frames within the time-out
period an internal retransmission counter is incremented by one,
if this counter exceeds the value of system parameter N2 the link
will be reset. If N2 has not been exceeded the value of the transmitter send state variable $V(S)$ is set to the value of the last valid
$N(R)$ received and the corresponding I-frames following this sequence
are re-transmitted with the Poll bit set on. The time-out condition
is only cleared by the receipt of a frame with its $N(R)$ advanced
and its Final bit set, any frames received with the Final bit not set
are ignored until the time-out condition has been terminated. When
a frame is retransmitted the timer T1 is restarted. If no frame with
the Final bit set on is received within the period of expiry of T1 the
frames are retransmitted up to N2 times. If no response is obtained
after N2 retransmissions the link is reset.

3.6 LINK SET-UP

Whenever the DCE is in an 'up' state and is able to proceed with a
Link level connection it will send a DISC frame to the DTE. This
DISC frame will be retransmitted every T1 seconds with the Poll bit
set on, as this is a re-transmission of the original DISC. When the
DTE wishes to reconnect to the DCE it waits until it receives a DISC
frame, it then replies with a UA within an interval of T1 seconds.
If the DCE does not receive a UA with the Final bit set on within
T1 it will continue in the down state by sending DISC. If a UA is
sent to the DCE it is then responsive to a link set-up command for a
period of T3 seconds. This link set-up command is the SABM frame.
If a SABM frame is not received by the DCE within interval T3 after
receiving the UA it will resume sending DISC, the initial DISC will
not have the Poll bit set but subsequent ones will. When the DCE
receives a SABM command from the DTE without the Poll bit set
it will reply with a UA without the Final bit being set. The link is

then in the 'up' state and I-frames may be sent in either direction within the constraints of the flow control which has been described.

3.7 LINK CLEAR-DOWN

To clear down a link which is in the up state the DTE sends a DISC frame without the Poll bit set. The DCE will respond by transmitting a UA frame, again without the Final bit set; it will then pause for T3 seconds. If the DTE wishes to bring the link back up immediately it may transmit a SABM frame after it receives the UA in response to its DISC. This must be received within T3 seconds of the receipt of the UA from the DCE.

Having described the function of the Link level protocol LAPB, some examples are now illustrated; the direction in which the frame is transmitted is indicated by the arrow which points either from the DCE to the DTE or vice versa.

Fig. 3.1 shows the simplest case of link set-up in which no frames go missing. The illustration assumes that the DTE has been down for some time hence the Poll bit is set on in the DISCs that are being sent as they are re-transmissions of the original DISC which was sent when the link-down state was entered.

	Direction	
DCE	⟷	DTE
DISC(P)	⟶	
interval T1		DTE in
DISC(P)	⟶	Down
interval T1		state
DISC(P)	⟶	
interval T1		
DISC(P)	⟶	DTE wakes up
	⟵	UA(F)
		T3 or less elapses
	⟵	SABM
UA	⟶	
		The link is now
		the up state

Fig. 3.1 – Link set-up.

| | Direction | |
DCE	$\longleftrightarrow$	DTE
I	$\longrightarrow$	Link in Up state
	$\longleftarrow$	I
	$\longleftarrow$	DISC
UA	$\longrightarrow$	Link cleared by DTE
interval T3		
DISC	$\longrightarrow$	Link in down state
interval T1		
DISC(P)	$\longrightarrow$	
interval T1		
DISC(P)	$\longrightarrow$	

Fig. 3.2 – Link clear-down by DTE.

Fig. 3.2 shows the simplest case of the clear-down of a link which was in the Up state followed by the DTE entering an unresponsive state. When the DTE becomes responsive again and wishes to re-establish the link it will follow the procedure illustrated in Fig. 3.1. If during either link set-up or clear-down frames are lost, or are ignored owing to FCS error, the normal protocol rules on re-transmission apply. If, for example, the UA(F) response by the DTE to the DISC(P) from the DCE is lost the SABM sent by the DTE is ignored and the DCE will continue to send DISC(P) at intervals of T1. If the SABM sent by the DTE to the DCE is lost or mutilated the DCE will fail to send the UA response within the T1 time-out period and will therefore re-transmit the SABM with the Poll bit set and will expect to receive a UA with the Final bit set on in reply. These conditions are shown in in Fig. 3.3.

We now give an example (Fig. 3.5) of the information transfer state showing the values of N(R) and N(S) as I-frames are exchanged along with RR to provide some of the acknowledgements. Notice in Fig. 3.5 that the acknowledgement from the receiver of a frame may either be sent by an explicit supervisory message (RR) or by an I-frame in the opposite direction if there is data being sent. Typically the transmitter software will use a timer to control the transmission of acknowledgements, it will not transmit an RR immediately if it

DCE	Direction ⟵⟶	DTE
DISC(P)	⟶	Link in Down state
DISC(P)	⟶	DTE wakes up
This UA(F) is lost	⟵	UA(F)
The SABM is ignored	⟵	SABM
DISC(P)	⟶	
	⟵	UA(F) DTE tries again
This SABM is lost	⟵	SABM interval T1 elapses
	⟵	SABM(P) re-transmission
UA(F)	⟶	Link in up state

Fig. 3.3 — Link set-up in which the initial UA(F) from the DTE is lost and then the SABM is lost.

DCE	Direction ⟵⟶	DTE
I	⟶	Link in Up state
	⟵	I
	⟵	DISC DTE clears link
UA	⟶	
		interval less than T3
	⟵	SABM DTE resets the link
UA	⟶	Link in Up state

Fig. 3.4 — Link clear-down by the DTE is followed by link set-up by the DTE within the interval T3.

has no data to be sent but will wait for a short interval to see if any data arrives from the higher levels to be sent before sending an RR.

Direction DCE DTE	Frame type	N(R)	N(S)
$\longrightarrow$	I	3	4
$\longleftarrow$	RR	5	
$\longrightarrow$	I	3	5
$\longleftarrow$	RR	6	
$\longleftarrow$	I	6	3
$\longleftarrow$	I	6	4
$\longrightarrow$	I	5	6
$\longleftarrow$	RR	7	
$\longrightarrow$	I	5	7
$\longleftarrow$	RR	0	
$\longleftarrow$	I	0	5
$\longleftarrow$	I	0	6
$\longrightarrow$	I	7	0
$\longleftarrow$	RR	1	

Fig. 3.5 – Typical Information transfer with
no re-transmissions or RNR states.

Fig. 3.6 shows the case of information transfer in which an I-frame
fails to reach its destination, the next I-frame which arrives therefore
has a sequence number which does not agree with the receiver's V(R)
state variable, in order to re-send the missing data frames it sends a
REJ command with the N(R) field set to the sequence number that
the receiver was expecting. Note that, although the frame was received
out of sequence, it was received without error, hence the N(R) value
which it carried will be valid if it is advanced from the previously
received value of N(R). Only one REJ is ever in operation in either
direction on a line at any time; any I-frame with an out-of-sequence
N(S) which arrives after a REJ has been sent is ignored.

The example shown in Fig. 3.6 is particularly simple as the data
was being sent from only one end of the link. If data had been being
sent from both ends then we might have seen the value of N(R) in
the I-frames which were re-transmitted altered from the values in the
original transmission of these frames. This can occur because, even
though the frame was received out of sequence, the FCS indicated
that this frame had not been corrupted and so the values of the
control fields were still valid. Fig. 3.7 shows an example of a more
complex case.

Direction

DCE DTE	Frame type	N(R)	N(S)	
←——	I	0	4	
←——	I	0	5	
←——	I	0	6	This frame is destroyed.
←——	I	0	7	This frame is out of order.
——→	REJ	6		Last two frames need resending
←——	I	0	6	
←——	I	0	7	
←——	I	0	0	
——→	RR	1		

Fig. 3.6 – Re-transmission of data after sequence error caused a REJ to be sent.

Direction

DCE DTE	Frame type	N(R)	N(S)	
——→	I	3	2	
——→	I	3	3	
←——	I	4	3	*
←——	I	4	4	This frame was destroyed.
←——	I	4	5	This frame is out of order.
——→	I	4	4	Acknowledges up to *.
——→	REJ	4		
←——	I	5	4	Acknowledges latest I-frame
←——	I	5	5	

Fig. 3.7 – An I-frame is sent before the receiver notices the out-of-sequence frame, a REJ is sent, the re-transmitted version of the frame contains an updated N(R) value to acknowledge the I-frame which was received.

An example is now given in which the DTE becomes flooded with data and is forced to issue a RNR to hold off further transmissions from the DCE. In Fig. 3.8 the DTE sends a RNR frame because its buffers have become full, the DCE responds by transmitting a supervisory command (i.e. RR, RNR, or REJ) with the Poll bit set on.

The supervisory frame is transmitted at intervals of T1 upto N2 times at which point the link would be reset. The DTE clears the busy condition by the transmission of either an RR or REJ frame with the Final bit set on, REJ is sent if the DTE wishes the DCE to re-send some of the data.

Direction DCE DTE	Frame type	N(R)	N(S)	
$\longrightarrow$	I	1	5	
$\longrightarrow$	I	1	6	
$\longrightarrow$	I	1	7	DTE now runs out of buffers
$\longleftarrow$ RNR		0		Acknowledges all I-frames.
interval T1 elapses				
$\longrightarrow$ RR(P)		0		DCE polls the DTE.
interval T1 elapses				
$\longrightarrow$ RR(P)		0		DCE polls again.
$\longleftarrow$ RR(F)		0		DTE no longer busy.
$\longrightarrow$	I	1	0	Next I-frame is sent.

Fig. 3.8 — DTE becomes busy for between two and three times T1 seconds; the DCE polls twice with RR(P) which elicits the RR(F) response from the DTE.

3.8 VALUES OF SYSTEM PARAMETERS IN SOME NETWORKS

Both here and in the section dealing with the X.25 Packet level inter-face we shall describe the implementation of the protocols described by four packet-switched networks. These are 'PSS' in the United Kingdom, 'Transpac' in France, 'Datex-P' in West Germany, and 'Telenet' operated by the GTE Telenet Communications Organisation in North America. We will list here the values assigned to the CCITT parameters referring to the link level.

The parameters dealt with here are:

    T1 — the re-transmission timer (milliseconds)
    N1 — the maximum number of bits in a frame
    N2 — the maximum number of re-transmissions
    K — the maximum number of unacknowledges frames.

Parameter	PSS	Transpac	Datex-P	Telenet
T1	100–10000	100–1600	3000	100–10000
N1	8232	1064	2104	8232
N2	20	10	10	3–50
K	7	1–7	1–7	7

Where a range of values is specified the values to be used are agreed with the network operator at subscription time. The parameters for other networks must be obtained directly from the network operators but are likely to be of the same order of magnitude as the ones shown. Most implementations of X.25 networking software have some means of configuration so that they may be run satisfactorily on any network.

4

The Packet level

4.1 INTRODUCTION TO THE PACKET LEVEL INTERFACE

This chapter is devoted to a detailed examination of the function and formats of the level 3 packet interface used in the X.25 protocol. It is the Packet level interface which characterises an X.25 network. In the CCITT specification the packets are transferred between the DTE and DCE by means of an HDLC link layer, the LAPB protocol we described previously. However, any transparent bidirectional virtual channel protocol could potentially be used to carry X.25 packets in a network because the user-visible characteristics are at the Packet level. In fact the whole X.25 protocol is concerned with the interface between the DTE (the network user), and the DCE (the network itself). No mention is made of how the network should operate internally. An X.25 network may consist of one or more packet-switiching exchanges which are interconnected in some way such that the exchanges can route calls from one DTE to another; they will not necessarily use X.25 between the nodes; they may not even work using virtual circuits but instead use an underlying data-gram network.

In X.25 each level 3 packet of information is transferred across the DCE/DTE interface inside a level 2 I-frame. Except for catastrophic events at the Link level the operation of level 2 has no effect on level 3. In order that a single connection to an X.25 network may carry more than one call at any time, every packet has a part of its header a 12-bit channel number which identifies which call this message relates to. Some administrations such as British Telecom choose to divide this into a four-bit logical group number and an eight-bit logical channel number, others simply refer to the full twelve-bit value. A DTE may at any time be conducting several calls with several different hosts, both Switched Virtual Calls (SVCs) and Permanent Virtual Circuits (PVCs), and the channel number serves to distinguish between them both for the DTE and the DCE. British Telecom's PSS network uses the four-bit group number to distinguish which of four types of call can be made on the logical channels in that group, the four types being:

Type	Group	Channels available	Complete channel number
PVC	0	1–255	1–255
	1	0–255	256–511
Incoming only –	2	0–255	512–767
SVC	3	0–255	768–1023
Both-way –	4	0–255	1024–1279
SVC	5	0–255	1280–1535
Outgoing only –	6	0–255	1536–1791
SVC	7	0–255	1792–2047

Channel number 0 is not a valid channel number in the CCITT scheme. PSS is not typical in this use of the logical group field to separate the different kinds of call that can be made, others such as the French Transpac network allow the user to define at sub-scription time which ranges of channel numbers are to be used for each of the four functions. Some X.25 networks do not support the distinction of three types of SVC and only provide Both-way type of SVC facilities. The terms Incoming, Both-way and Outgoing refer in this context to the ability of the DTE to set up calls. In the case of Outgoing only channels the DTE may choose to make calls out on those channels but the network will never send calls in to the DTE on those channels. Conversely the network may make calls into a DTE on Incoming only channels but the DTE is not permitted to make outgoing calls on them. Both-way channels permit either the DTE or the network to make calls out or in on them. This facility if typically used to prevent the flooding of some system by either internal or external factors. A host might be configured typically for two Incoming only, twelve Both-way, and two Outgoing only channels, thus if fifteen remote calls are set up to that host the fifteenth one will fail as the network will not be able to find a free channel on which to make an incoming call to the DTE, the DTE however will still be able to make two outgoing calls because it can use the channels reserved for that purpose.

To minimise on call clashes where both the DTE and DCE attempt to use the same channel number at the same time to make a call, the CCITT recommend that DTEs use the highest free channel number available to make outgoing calls and the DCE uses the lowest free channel number to make incoming calls. This means that call clashes will only occur when all but one of the channels allocated to a DTE have been used up. This explains the reason for the order of group numbers given in the PSS scheme above. For networks which do not pre-allocate group numbers for this purpose, the order of Incoming, Both-way, and Outgoing channels is always the same as used for PSS for the same reason as given above.

A Permanent Virtual Circuit provides the same facilities as a Switched Virtual Call, except that the end-to-end connection is always present as long as both DTEs are running – no call set-up phase is required. PVCs are defined by the subscriber to the PTT running the network at subscription time, the PTT will specify a channel number on which that PVC will exist. An SVC by contrast is set up dynamically on one of the channels which the PTT has agreed that SVCs may use and will only exist until it is cleared by one DTE or the other, or one of the DTEs stops running its communications link.

4.2 STARTING THE PACKET LEVEL INTERFACE

The restart procedure is used to ensure that both the DCE and DTE consider that all the PVCs and SVCs are in a known state. A restart causes all PVCs to be reset and all SVCs to be cleared and free to be used. The Packet level will need to be restarted if the Link level has failed for some reason, for example the line between the DTE and DCE was disconnected. If the Link level has timed out and has been restarted by the exchange of SABM, UA, etc., as previously described, the Packet level will be restarted by the DCE sending to the DTE a Packet level Restart Indication frame; in PSS the Restart Indication frame will be sent to the DTE at intervals of six seconds until a Restart Confirmation is received. The DTE may at any time transmit a Restart Indication to the DCE if it needs to restart the Packet level completely, for example if it has completely lost track of the state of the active SVCs. A Clear is sent to all SVCs by the network and the Restart Confirmation is not sent to the DTE until Clear Confirmations have been received on all previously active SVCs. Thus the speed of the restart is limited by the speed of the slowest remote DTE to respond to the Clear Request. To prevent a faulty DTE from indefinitely holding up a restart the DCE will send a Restart Confirmation after ninety seconds even if Clear Confirmations have not been received from all the remote DTEs with which SVCs had been established.

The format of the level 3 restart packet is slightly unusual in that, although the group and channel number fields are present, they both contain zeros, obviously because these fields are irrelevant. The general format of a level 3 packet is:

Bits:	8	7	6	5	4	3	2	1
Oct 1		GFI				LGN		
2				LCN				
3				PTI				

4 onwards — type specified fields

where GFI = General Format Identifier, normally 0001
       LGN = Logical Group Number
       LCN = Logical Channel Number
       PTI = Packet Type Identifier

The format of a level 3 Restart Indication packet is:

Bits:	8	7	6	5	4	3	2	1	
Oct 1	0	0	0	1	0	0	0	0	GFI and zero LGN
2	0	0	0	0	0	0	0	0	Zero LCN
3	1	1	1	1	1	0	1	1	PTI
4				Restarting Cause					
5				Diagnostic Code					Optional

The Restart Indication packet contains a Restart Cause field which is defined to have three possible values: 1 which implies a local procedure error, 3 for network congestion, and 7 for network operational. The first would normally be the reason for a DTE initiated restart, while the last is the usual reason for a restart when the network has recovered from some failure.

The recipient of a Packet level Restart Indication packet should reply with a Packet level restart confirmation. This takes the following format:

Bits:	8	7	6	5	4	3	2	1	
Oct 1	0	0	0	1	0	0	0	0	GFI and zero LGN
2	0	0	0	0	0	0	0	0	Zero LCN
3	1	1	1	1	1	1	1	1	PTI

The Diagnostic field of the restart indication packet is passed in the Diagnostic field of the Reset or Clear packets that the restart causes to any PVCs or SVCs active at the time of the restart. The values for the Diagnostic field are defined by the operator of the network.

4.3 ESTABLISHING A VIRTUAL CALL

The typical traffic handled by a packet-switched network is of the Switched Virtual Call (SVC) type; Permanent Virtual Circuits (PVCs) are only occasionally used, often for network management between nodes in a network. When an SVC is to be established the DTE will send a Call Request type packet. Calls are only ever initiated by a DTE, never by the network. The Call Request packet contains the address of the destination DTE to which the call is being made. This address is a decimal number up to twelve digits long with an optional two-digit subaddress which may be used to identify the service that is being requested on the destination DTE. This has only end-to-end significance, its use being bilaterally agreed in advance. The DTE making the Call Request will specify a logical group and channel number in the packet which will serve in all future interactions with the network to identify the conversation. By convention the DTE will choose the highest free channel number from the range on which it is permitted to make outgoing calls, as was mentioned in section 4.1, in order to minimise the risk of a call collision occurring. The Call Request packet may include the address of the DTE making the call, however, this field is optional since the network will insert this value when the Call Request is passed to the destined DTE as an Incoming Call packet. The Called and Calling addresses are sent as binary coded decimal values packed two to each octet. The lengths of the addresses precede the addresses as two four-bit value in a single octet. The addresses are followed by a variable length Facility field and a sixteen-octet optional user data field. We shall later discuss an alternative format which allows up to 128 octets of user data, termed the Extended Format.

The format of a Call Request packet, using the same notation as before is:

Bits:	8	7	6	5	4	3	2	1	
Oct 1	0	0	0	1		LGN			GFI and LGN
2				LCN					LCN
3	0	0	0	0	1	0	1	1	PTI
4	Calling length				Called length				Address length indicator
5		Up to 14 octets addresses							Padded to integral no. of octets
n	0	0		Facility field length					Next octet after addresses
		Optional Facility fields							Next octets are facilities
m		Call user data field							Optional up to 16 octets

Notice that after the initial four octets the format is variable: two addresses of up to 14 decimal digits each are followed by a length field whose value is in the range zero to 63 to show the length of the optional Facilities field. The Facilities field is used to convey information about the call to the network by the DTE. The Incoming Call packet has exactly the same format, if the calling address was not filled in by the caller it is inserted by the network, the facilities field may contain information from the network for the destined DTE. If a particular facility has been requested by the caller the network may alter the values, the called DTE may send altered values back in the Extended Format type of Call Accept packet. The called DTE may reply to an Incoming Call packet in one of two ways — either by a Call Accepted packet which is returned to the caller as a Call Connected packet, or by sending a Call Cleared packet to reject the call request. The format of a Call Accepted packet is now given:

Bits:	8	7	6	5	4	3	2	1	
Oct 1	0	0	0	1		LGN			GFI and LGN
2				LCN					LCN
3	0	0	0	0	1	1	1	1	PTI

This format is also used as the format of the Call Connected response to the Calling DTE.

The Clear Request packet type is used to reject unwanted incoming call requests and also to clear down existing SVCs which are no longer needed. In the standard forms one octet of data can be passed from the clearer by means of a diagnostic octet, a Cause octet is also included which is always set to zero if the call was cleared by one of the DTEs involved in a call. If the network rather than one of the DTEs clears the call a value is placed in the Cause octet to indicate the reason for the network having cleared the call; these value are standardised by CCITT and added to by the network operators. The format of the Clear Request is now shown:

Bits:	8	7	6	5	4	3	2	1	
Oct 1	0	0	0	1		LGN			GFI and LGN
2				LCN					LCN
3	0	0	0	1	0	0	1	1	PTI
4	0	0	0	0	0	0	0	0	Cause field always zero
5			Diagnostic Code						Diagnostic octet

The Clear Request packet will produce a response to the clearing DTE when the call has been cleared in the form of a Clear Confirmation packet. Thus the DTE wishing to clear the call or refuse an incoming Call Request will send a Clear Request which will be delivered to the destination DTE as a Clear Indication packet with exactly the same format as a Clear Request packet, the DTE receiving the Clear Indication will then respond with a Clear Confirmation packet which will be delivered to the clearer as an incoming Clear Confirmation packet. The format of the Clear Confirmation packet is:

Bits:	8	7	6	5	4	3	2	1	
Oct 1	0	0	0	1		LGN			GFI and LGN
2				LCN					LCN
3	0	0	0	1	0	1	1	1	PTI

The formats and uses of the Extended Format packet types will now be described.

4.4 EXTENDED FORMAT PACKET TYPES

The Call Request and Call Accept which were described in the previous

section were the standard X.25 packet sequences for the establishment of SVCs, an optional extension permits a DTE to add extra
information to the call set-up stage. These facilities are not mandatory
and are not implemented by all network vendors. The facilities
provided allow the receiver of an incoming Call Request packet to
negotiate various parameters to optimise the data transfer for his
application, and also to send up to 128 octets of user data with the
Call Request, Call Accepted and Call Cleared packets. These two
facilities are termed generically the Extended Format and Fast Select
formats.

The Extended Format Call Accepted packet allows the DTE on
receipt of an incoming Call Request to negotiate the Transmit and
Receive window sizes to be used for packet level flow control, and
also the packet size to be used in the transfer of Data packets. The
packet sizes may be different in each direction of the SVC; however,
those networks which do implement Extended Format facilities
generally insist that the packet sizes and window sizes are the same
in both directions. The negotiation in Extended Format Call Accept
must be in the downward direction otherwise the call will be cleared
with 'Procedure Error' as the cause. Typical default sizes for the
window and packet size are 2 and 128 octets respectively. The
need for a larger window size increases with increased line speed, a
window of 2 is satisfactory for 2400 bit/second line speeds, for 4800
bits/second a window size of 4 is optimal and for speeds of 9600
bits/second the full window of 7 is recommended. The Extended
Format Call Accepted packet is similar to the ordinary Call Accepted
packet with the addition of an Address and Facilities field. The
format of the Extended Call Accepted packet is now given:

Bits:	8	7	6	5	4	3	2	1	
Oct 1	0	0	0	1	LGN				GFI and LGN
2	LCN								LCN
3	0	0	0	0	1	1	1	1	PTI
4	Calling length				Called length				Address length indicator
5	Up to 14 octets addresses								Padded to integral no. of octets
n	0	0	Facilities field length						Next octet after addresses
m	Optional Facilities fields								Up to 63 octets of facilities

This packet format is exactly repeated as an incoming Call Connected packet to the initiator of an SVC connection, the network will have filled in the Address fields of the packet which are not mandatory for the Call Accept packet from the called DTE. Either or both of the Address length fields may be zero and the Address fields which follow may therefore empty. As in the Call Request the addresses are in binary coded decimal up to 14 octets in total with two characters packed in each octet, if the total number of characters is odd the final octet is padded with zeros to make an integral number of octets in the address. The Facilities field is up to 63 octets in length since the length field is only six bits wide, it may also be null. The contents of the Facilities field is the same for Call Request, Extended Call Accepted and Fast Select forms of packet and will be described after the Fast Select formats have been described. In an incoming Call Connected packet with extended format the Facilities field will contain the actual values for the window and packet sizes to be used for this connection even if the Call Accepted packet did not make specific mention of the values and left the default values.

The Fast Select formats of Call Request, Accept and Clear allow for the transmission of up to 128 octets of user data from end to end with them. In certain circumstances this is a very useful addition. The two main uses of these formats are the provision of datagram-like facilities for very short types of transactions and the use of multi-domain addressing where a call has to traverse more than one network to reach its destination. Two types of Fast Select Call Request exist, termed 'Restricted Response' and 'Unrestricted Response'. In the former the only legal response is a Clear Request (either normal or Fast Select format), in the latter case either a Call Accepted (either normal, Extended or Fast Select format) or a Call Cleared (normal or Fast Select). The only circumstances in which a Fast Select Call Cleared packet may be sent is in response to an Incoming Call as described above; it may not be sent in response to Data. A normal Call Request packet may be replied to with either a normal Call Accept or Call Cleared or by an Extended format Call Accepted packet, but not by a Fast Select format packet.

The format of the Fast Select Call Request packet is identical to that of the normal format except that the Call User Data Field at the end of the packet may be up to 128 octets in length, the Facilities field is mandatory and must include the codes to inform the destination DTE whether it is a 'Restricted Response' or an 'Unrestricted Response' Call Request, all other facilities are optional. The Fast Select format Incoming Call packet has a format identical with the Call Request packet except that the Address fields may have been completed if they were omitted by the Calling DTE. The format of the Fast Select Call Accepted packet is the same as the Extended Format Call Accepted packet, with the addition of a Call User Data Field of up to 128 octets at the end. The Facilities field may only contain packet size and window size codes if the Incoming Call packet contained values for these parameters in its Facilities field. If they are included where the Incoming Call did not contain them it is considered to be a procedure error and the call will be cleared by the network. Any other parameters found in the Facilities field will be ignored and not passed back to the calling DTE in the Call Connected packet which it will receive. The Fast Select format of the Call Connected packet is identical with that of the Fast Select Call Accepted packet just described. As before with Call Connected packets, the network will fill in the values of the Called and Calling addresses if they are not included by the called DTE in the Call Accepted packet. If the DTE which made the Call Request subscribes to the Fast Select option then it will always receive Fast Select formats of Call Connected and Incoming Call packets. If the remote DTE does not reply with a Fast Select format packet the Fast Select packet format is generated by the network and will contain default values in the Facilities field for the window and packet sizes and a zero length User Data field. The format of the Fast Select format of Clear Request is now given:

Bits:	8	7	6	5	4	3	2	1	
Oct 1	0	0	0	1		LGN			GFI and LGN
2				LCN					LCN
3	0	0	0	1	0	0	1	1	PTI
4				Clearing cause					
5				Diagnostic code					User defined
6	0	0	0	0	0	0	0	0	Zero length Address field
7	0	0	0	0	0	0	0	0	Zero length Facilities field
8				Start of user data					Up to 128 octets of user data

When the Clear Indication is received by the remote DTE it will be identical to the Fast Select Clear Request sent except that the network will have inserted the called and calling address into the Address fields of the packet, padded with a null semi-octet if the length is an odd number of digits.

The two main uses of the Fast Select facilities are for increased addressing range and for datagram facilities for small transactions. The fourteen-digit addresses of the X.25 protocol are quite restricting, while it is sufficient for a single network there are serious problems where two or more networks are linked together. This situation is common: many organisations operate their own internal networks and also have a requirement to interconnect with other organisations' networks, possibly using a public packet-switched network to provide the link between them. In the UK academic community, for example, most universities have their own X.25 network, these can interwork using either PSS, the national packet-switched network provided by British Telecom, or by JANET — the Joint Academic Network. The use of either of the two intervening networks involves three addressing domains which must be provided with addressing information. In this case the Transport protocol makes use of the 128 octets of User Data in the Fast Select format of Call Request to pass further addressing information forward from one network to the next; we shall describe this further in Chapter 5 where the various Transport protocols in use are described. A typical use of the datagram facility is the increasing use of Point of Sales (POS) terminals which are able to perform Electronic Funds Transfer (EFT) immediately a transaction is made. The amount of information to be transferred is small and the setting up and clearing down of a complete SVC would be wasteful. The information to be transferred is typically the account number of the account to debited, and the amount to be transferred; the security of the purchaser may be improved by requiring him to enter a Personnal Identification Number (PIN) when the transaction is made. This information may possibly be encrypted to provide further security to the banks etc. involved in the transaction, the response will be either 'yes' or 'no' depending whether the purchaser has enough funds to cover the purchase, and possibly a sequence number to identify the transaction. This sort of information can easily be passed in a Fast Select Call Request with Restricted Response such that the only possible reply by the bank is a Fast Select Clear Request with User Data in both directions carrying the information concerning the transaction.

4.5 FACILITY FIELD CODING

Both the Call Request and Call Accept packet have a field of variable length up to 63 octets for carrying requests for special facilities. The first octet of each facility request, there may be several in the Facilities field, contains the class of the facility and a code which identifies the facility. There are four classes of facility termed A, B, C, D. Classes

A, B, and C have one, two and three octets of information following the header octet, class D facilities have a length field following the header octet which contains the length of the parameters which follow. The general format of a class A facility is:

Bits:	8	7	6	5	4	3	2	1
Oct 1	0	0		Facility code					Class and Facility code
2	One octet Facility parameter							

The format of class B facilities fields is:

Bits:	8	7	6	5	4	3	2	1
Oct 1	0	1		Facility code					Class and Facility code
2	First parameter octet							
3	Second parameter octet							

The format of class C facilities fields is:

Bits.	8	7	6	5	4	3	2	1
Oct 1	1	0		Facility code					Class and Facility code
2	First parameter octet							
3	Second parameter octet							
4	Third parameter octet							

The format of class D facilities fields is:

Bits:	8	7	6	5	4	3	2	1
Oct 1	1	1		Facility code					Class and Facility code
2	Binary coded decimal length								Value between 4 and 9
3	Between 3 and 9 octets of							
n	facility parameters							

The coding and usage of the facilities offered by British Telecom's PSS service are now described as a representative sample of the facilities offered. The facilities to be described are standard parts of the X.25 protocol, there is also a National Options marker which can be used to introduce facilities which are strictly local to the requirements of the network vendor and his customers. Reverse charging is a facility which uses class A type facilities parameters; it provides a facility analogous with 'collect' calls in the ordinary telephone system. This facility would be inserted in the Call Request packet of a DTE calling another host, the receiving host would notice the Reverse Charging facility field in the Incoming Call packet and decide either to accept or reject the call. If the network provider offers a special discount rate of charging for bulk it may be economic where a central system is accessed by many satellite systems for the remote systems to make reverse-charged calls to the central system so that it qualifies for the bulk rate. The Fast Select option which we mentioned in the previous section is also a class A facility, as is the National Options marker and the Closed User Group (CUG) facility. A Closed User Group is a set of DTEs which at subscription time request to be placed in a CUG, the effect is to make the DTEs appear to be in a network of their own. The CUG is identified by a two-digit number agreed with the network provider at subscription time. A DTE may be a member of several CUGs, the CUG to be used in a particular call is identified by the number in the facility parameter in the Call Request. There are three types of CUG, those which permit outgoing access to any destination as well as to other members of the GUG, those which permit incoming access by other DTEs to members of the CUG but which allows members of the CUG only to be able to call other CUG members, and Bilateral Closed User Groups which can both access and be accessed by outside DTEs as well as accessing other CUG members. The encodings of these class A facilities are now given:

The facility code for Reverse Charging, Fast Select with restricted response and Fast Select with unrestricted response is the same:

Bits:	8	7	6	5	4	3	2	1	
Oct 1	0	0	0	0	0	0	0	1	Class A and Facility code
Oct 2			See below						

Octet 2 encodings:

Bits:	8	7	6	5	4	3	2	1	
	X	X	0	0	0	0	0	1	Reverse Charging, XX ignored
	1	1	0	0	0	0	0	Y	Fast Select — restricted response
	1	0	0	0	0	0	0	Y	Fast Select, Y ignored

Closed User Group access uses the following Facilities Code:

Bits:	8	7	6	5	4	3	2	1	
Oct 1	0	0	0	0	0	0	1	1	Class A and Facility code
2	First digit				Second digit				Two-digit CUG number

Note that the digits are binary coded decimal values from zero to nine.

The codings for class B facilities are now given.

The Window Size negotiation facility uses class B parameters in order to indicate what size of window should be used in each direction, i.e. to and from the DTE. The values are given as binary numbers in the range one to seven. The format is:

Bits:	8	7	6	5	4	3	2	1	
Oct 1	0	1	0	0	0	0	1	1	Class B and Facility code
2	0	0	0	0	To DTE				Binary Window Size to DTE
3	0	0	0	0	From DTE				Binary Window Size from DTE

The default window size is two in British Telecom's PSS network. This value is negotiated on a per call basis, by inserting the facility in the Call Request packet sent out by a DTE. The receiving DTE will see this value in its Incoming Call packet and may either accept the call with the parameters as given or downwards negotiate the values of the window sizes if insufficient buffers are available to use the window size requested. In PSS the values of the window size must be identical in both directions, but in other networks this may not necessarily be the case. The use of different window sizes in each direction may reflect different throughput requirements in the two directions. In general superior performance is obtained by the use of the largest possible window size, especially where the speed of the line is high. The functioning of the window at the packet level will be described shortly in section 4.6. The other influence on through-put, apart from the window size, is the maximum packet size, which may be in the range 128 to 1024 octets of user data on the PSS network, other networks allow the maximum packet size to be dropped as low as 64 octets or up to 2048 octets. The packet size facility is operated in the same way as window size in that the initiator of a call may request a particular maximum packet size to be used, the remote DTE can downwards negotiate the value if it cannot cope with the value given. The coding of the Packet Size facility is now given:

Bits:	8	7	6	5	4	3	2	1	
Oct 1	0	1	0	0	0	0	1	0	Class B and Facility code
2	0	0	0	0	To DTE				Binary log. of Packet Size to DTE
4	0	0	0	0	From DTE				Binary log. of Packet Size from DTE

As in the case of window size negotiation the packet sizes can be different to and from the DTE making the call, again PSS requires these two values to be the same. The values used are the binary logarithms to base 2 of the packet size, e.g. the default value of 128 is coded as binary 0111 or 2 to the power 7.

PSS at present makes no use of any class C facilities, there are two class D facilities in use: the Call Duration facility and the Segment Count facility, these are both returned in the Fast Select Clear Indication packet, this is identical to the Fast Select Clear Request packet except that the network inserts a Facilities field as well as passing through the User Data field. The Call Duration and Segment Count facilities are subscription time options, if a DTE subscribes to the facility then all incoming Clear Indication packets will be of the Fast Select type with the duration and count facilities included except if the Call Request was of the restricted response type in which case a fixed segment count of six is always incurred for accounting purposes. The format of the Call Duration facility is now given:

Bits:	8	7	6	5	4	3	2	1	
Oct 1	1	1	0	0	0	0	0	1	Class D and Facility code
2	0	0	0	0	0	1	0	0	Length count field of 4
3	Number of days								Binary number of days
4	Number of hours								Binary number of hours
5	Number of minutes								Binary number of minutes
6	Number of seconds								Binary number of seconds

Note that in class D facilities the second octet indicates the number of octets which will follow rather than the total number of octets in the facility. In the case of the Segment Count facility the number of octets following the two header octets is eight. In this case the number of octets transferred in each direction is given as a 32-bit integer with the most significant octet first in each case. The coding of the Segment Count is now given:

Bits:	8	7	6	5	4	3	2	1	
Oct 1	1	1	0	0	0	0	1	0	Class D and Facility code
2	0	0	0	0	1	0	0	0	Length count field of 8
3	Most significant octet of								
4	Received Segment Count								Received segments
5									
6	Least significant octet								
7	Most significant octet of								
8	Sent Segment Count								Sent segments
9									
10	Least significant octet								

Call Duration and Segment Count facilities are often used by gateways which allow access to users of expensive networks such as IPSS (International Packet Switched Stream) by British Telecom, where access to a host in another continent can cost many pounds sterling per hour of connection or per thousand segments sent, a segment being 64 octets or a part of 64 octets if a partly empty packet is sent. Comparisons are performed between the locally calculated cost and that billed by the network provider. In the case of PSS and IPSS this comparison is particularly easy to make as one of the nationally defined extensions allows the cost of a call to be returned as part of a Fast Select Clear Indication packet.

4.6 DATA TRANSFER PACKETS

The data transfer phase of either a PVC or an SVC makes use of four packet types, Data, Receiver Ready (RR), Receiver Not Ready (RNR), and Reject (REJ). These packets, used in conjunction with each other, allow the receiver of a stream of data to control the speed at which it is sent to it by the network; this is performed by the use of 'flow control' which will be described after the basic formats of the packets have been given. As the name implies the Data packet is the packet type used to send user data from one DTE across the network to another DTE. The format of the Data packet is now given:

Bits:	8	7	6	5	4	3	2	1	
Oct 1	Q	0	0	1		LGN			Qualifier, GFI and LGN
2				LCN					LCN
3		P(R)		M		P(S)		0	Sequence counts and More bit
4				User data					0–128 octets of user data

The Data packet above is shown using the same notation as before. Up to 128 octets of user data may follow the three-octet header, or up to 2048 if the packet size has been successfully negotiated to this size. The X.25 protocol is slightly inconsistent in its use of the packet header for data packets in that the General Format Identifier is used to carry the 'Qualifier' bit which is a user-settable bit which has significance from end to end only. This bit is typically used by a higher level protocol to differentiate between data which is intended for a user process and data which forms part of the high level protocol itself. The other user-defined bit is the 'More' bit, this is used to indicate that a packet is part of a series of packets which together make up a single logical message. The More bit may only be set on if the packet is full, i.e. contains the same number of octets of user data as was negotiated at call set-up time; if a partially filled packet is sent with the More bit set on the network will clear the call since a procedure error will have been detected. If the More bit is not set on, the packet may contain any number of data octets, some networks, however, prohibit the transmission of packets with zero octets of user data. The two sequence counts in octet three are used to provide the flow control mechanism as we shall now describe. It should be remembered that flow control at the packet level only has significance for the SVC or PVC using the Logical Group and Channel number given in the header octets, stopping the flow of data on one virtual circuit does not affect the flow of data on any other. This property should be considered in comparison with the flow control exerted at the Link level where the blocking of further transmission will halt the flow of data for all calls in progress. It is obviously mandatory to permit users of X.25 packet level interfaces only to control data at the packet level, so as to avoid interference with other users. The flow control mechanism used is very

similar to that employed by LAPB, except that here the fact that the
Packet level makes use of a virtual error-free channel provided by
the Link level can be used to render unnecessary any error detection
or correction procedures. The other packet types involved with the
transmission of data are now detailed. The Receiver Ready (RR) is
shown first:

Bits:	8	7	6	5	4	3	2	1	
Oct 1	0	0	0	1		LGN			GFI and LGN
2				LCN					LCN
3		P(R)		| 0	0	0	0	1	Sequence count and PTI

The Receiver Not Ready (RNR) packet:

Bits:	8	7	6	5	4	3	2	1	
Oct 1	0	0	0	1		LGN			GFI and LGN
2				LCN					LCN
3		P(R)		| 0	0	1	0	1	Sequence count and PTI

The Reject (REJ) packet:

Bits:	8	7	6	5	4	3	2	1	
Oct 1	0	0	0	1		LGN			GFI and LGN
2				LCN					LCN
3		P(R)		| 0	1	0	0	1	Sequence count and PTI

As will be recalled from the previous section on the negotiation
of facilities, the range of window sizes used at the Packet level of
X.25 is the same as at the Link level, i.e. up to seven. It can be seen
seen from the formats of the Data, RR, RNR, and REJ packets that
they all contain three-bit counter fields, termed $P(R)$ and $P(S)$,
which can count between zero and seven. The $P(R)$ field is termed
the Received sequence count and the $P(S)$ the Sent sequence count.
Each time a new packet is sent to the Link level for transport to the
network node to which the DTE is connected, the Sent sequence
count is incremented, except that if its current value is seven it will
be set back to zero. If the difference between the last Received

sequence count and the Sent sequence count is greater than the value agreed for this virtual circuit, then the data is not sent and the flow in that direction is blocked until a packet is received which updates the Received sequence count. The Received sequence count can be sent on any of the four data transfer type packets, i.e. Data, Receiver Ready, Receiver Not Ready or Reject. If data is flowing in both directions at a rate which both ends of the link can deal with satisfatorily then no other packet types need to be sent, if there is no data to be sent outwards from a DTE which is receiving Data packets from some remote host then either the Receiver Ready or Receiver Not Ready packet type may be used to carry the latest Received sequence count. The RR type packet is used to indicate that the DTE has sufficient buffer space and resources to continue to receive packets of data, the RNR type packet would be used to inform the remote DTE that, owing to resource constraints, no further packets of data should be forwarded on this virtual channel until a Receiver Ready packet is sent to indicate that the congestion situation is over. It may be that a DTE receiving data will run out of buffer space for further packets and send a Receiver Not Ready packet at the same time as further Data packets are being sent to it, the link protocol being truly full duplex in nature. If this situation occurs and the receiving DTE is forced to throw away a packet of data any subsequent packets would have a sequence count which is not in sequence. To indicate that packets need to be retransmitted the Reject packet type is used, in the same way as at Level 2, to indicate the next sequence count it expects to receive from the network; thus the remote DTE must restransmit this packet and any it has been subsequently. The sequence counters both start at zero when the call is set-up, and are reset to zero should the call be Reset, a procedure which will be decribed shortly. The response to a data packet transmitted to the network, i.e. the reception of a $P(R)$ greater than or equal to the $P(S)$ of the Data packet, must be received within a certain time, termed the Level 3 time-out period; failure to receive the necessary confirmation of receipt will cause the call to be reset. Hence the RNR packet should be used to hold off data rather than indefinitely holding back the acknowledgement.

We will now show a session during which a call is received by a DTE, a Call Accept is sent followed by the transfer of data and clear-down. This example will first illustrate only the Packet level frame transfer, but will then be repeated to show the operation of both the Link and Packet levels for this transaction. The transaction shown is that of an interactive terminal calling a host computer using a Packet Assembler/Disassembler (PAD) using the X.29 protocol which will be described om Chapter 6. This protocol makes use of the Qualified data bit to signify messages destined for the control of the PAD itself rather than for the terminal. After the connection has been established the host system sends a message to the PAD to set it into a mode suitable for working with this machine's operating system, the PAD replies showing the parameters set up as the host requested; the host then resets the PADs mode this time not asking for a reply. The remaining data is not Qualified and is destined for the terminal or host, depending on the direction, without alteration. All the frames have the same logical group and channel number so this is not shown here.

Direction	Q bit	Packet type	Received P(R)	P(S)	Transmitted P(R)	P(S)
Rx→		Call Req.				
←Tx		Call Acc.				
←Tx	Q	Data			0	0
Rx→	Q	Data	1	0		
←tx		RR			1	
←Tx	Q	Data			1	1
←Tx		Data			1	2
Rx→		RR	2			
Rx→		RR	3			
Rx→		Data	3	1		
←Tx		RR			2	
←Tx		Data			2	3
←Tx		Data			2	4
Rx→		RR	4			
Rx→		RR	5			
Rx→		Clr Req.				
←Tx		Clr Conf.				

It can be seen from this example that the holdback timer on the DTE, which controls how long the DTE should wait to see if a Data packet from the system itself comes along to be sent out rather than sending out an RR frame when an acknowledgement is required, is set to a short interval so that the Data packets rarely carry acknowledgements with them. The adjustment of the holdback timer is generally a compromise between the *Scylla* of reduced response time and the *Charybdis* of increased utilisation of the channel by packets which are strictly redundant. For a host handling mostly interactive work, the increased response of immediate acknowledgment may be advantageous, whereas for a host handling mostly bulk data transfer the increased overall throughput of the line will probably be of greater importance. The same set of Packet level transactions shown above is now shown but with the operation of the Link level also given:

Direction	Frame type	Received N(R)	Received N(S)	Transmitted N(R)	Transmitted N(S)	Q bit	Packet type	Received P(R)	Received P(S)	Transmitted P(R)	Transmitted P(S)
Rx→	I	1	5				Call Req.				
←Tx	RR			6							
←Tx	I			6	1		Call Acc.				
Rx→	RR	2									
←Tx	I			6	2	Q	Data			0	0
Rx→	I	3	6			Q	Data	1	0		
←Tx	RR			7							
←Tx	I			7	3		RR			1	
←Tx	I			7	4	Q	Data			1	1
←Tx	I			7	5		Data			1	2
Rx→	I	6	7				RR	2			
←Tx	RR			0							
Rx→	I	6	0				RR	3			
←Tx	RR			1							
←Tx	RR(P)			1							
Rx→	RR(F)	6									
Rx→	I	6	1				Data	3	1		
←Tx	RR			2							
←Tx	I			2	6		RR			2	
←Tx	I			2	7		Data			2	3
←Tx	I			2	0		Data			2	4
Rx→	I	1	2				RR	4			
←Tx	RR			3							
Rx→	I	1	3				RR	5			
Rx→	I	1	4				Clr Req.				
←Tx	I			5	1		Clr Conf.				

In the preceding example the similarity between the packet and link level is highlighted. This is a rather simple example in which only a single virtual call is operating at the Packet level, obviously it is much harder to follow when several calls are proceeding at the same time. With the aid of a reasonably intelligent line-monitor it is possible to watch the progress of the Packet and Link level in real-time, which will give a better idea of the timings involved. On a two-dimensional diagram it is difficult to show the occurrence of time-outs etc. and the real flavour of the protocol is best seen as it happens in a live situation. The above example shows the difference between the Link level and Packet level holdback timers, the piggy-backing of acknowledgements at the Link level is quite common in this example.

Having shown the basic data transfer mechanism at the Packet level the other facilities are now described, such as Call Reset and Interrupt data.

4.7 INTERRUPT PACKETS

A frequent requirement when communicating between an interactive terminal and a host computer is the provision of a facility to allow the user to inform the computer that it should stop what it is at present doing and allow the user to regain control of the system. Depending on the computer system in use it may or may not be possible to inform the host machine of such a request in the ordinary flow of data from the terminal to the host. In the situation where the host has stopped receiving input for flow control reasons it would be impossible to receive the user's request for attention. This problem is overcome by the provision in the X.25 protocol of an Interrupt facility which operates outside the normal flow control constraints of Data packet transmissions. Since this means of sending data has no flow control in it, a restriction is imposed that a DTE cannot have more than one outstanding Interrupt on any virtual call in each direction. Thus the receiving software need only reserve one byte of buffer space for each logical channel in use in case an Interrupt is received. Higher level protocols for handling interactive terminals generally make use of the Interrupt facility to provide a standardised way of causing an attention signal to be raised in a host. The format of the Interrupt packet is now given:

Bits:	8	7	6	5	4	3	2	1	
Oct 1	0	0	0	1		LGN			GFI and LGN
2				LCN					LCN
3	0	0	1	0	0	0	1	1	PTI
4				User data					

The format of the Interrupt packet is identical for the receiver
except that the Logical Group and Channel numbers will have been
altered in the usual manner by the network to reflect the correct
Channel numbers for this end of the link. If more than one octet
of user data is inserted into an Interrupt packet the call will be reset
owing to the protocol error. The 1984 version of X.25 will permit
up to 32 octets of user data in an Interrupt packet. The Interrupt
packet is not actually used to carry any user data in the most popular
terminal handling protocol − X.29 − the Interrupt packet alone
signifies to the receiver that attention is being sought by the terminal.
The actual amount of data which the definition of X.25 allows to
be passed by an Interrupt packet is not of great importance since
there will never be more than one Interrupt packet outstanding in
each direction of a link, the software need only reserve enough
storage for one Interrupt per virtual circuit in use. The receiver of
an Interrupt should respond as soon as is possible with an Interrupt
confirmation packet which will be sent by the network through to
the originator of the Interrupt packet. After the network has sent
an Interrupt Confirmation packet to a DTE it is then willing to
receive another Interrupt packet. The network will reset the call
if an Interrupt is sent by a DTE before it has received the Confirmation packet for a previous Interrupt packet. The format of the
Interrupt Confirmation packet is now given:

Bits:	8	7	6	5	4	3	2	1	
Oct 1	0	0	0	1		LGN			GFI and LGN
2				LCN					LCN
3	0	0	1	0	0	1	1	1	PTI

Note that there is no User Data field on an Interrupt Confirmation
packet.

4.8 CALL RESET

If the network detects a protocol error by one of the DTEs involved
in a virtual circuit, or if one of the DTEs wishes to throw away any
data in transit, or buffered at either end of the link, in order to
return the link to a known state, a Reset is performed. If the network
generates the Reset, the Reset Indication packets will be sent to the
DTEs at both ends of the link, the Reset operation is complete when
each of the DTEs have returned a Reset Confirmation to the network.
If the Reset was generated by one of the DTEs it will send a Reset

Request to the network, the remote DTE will deceive notification of the Reset in a Reset Indication packet; it must then reply with a Reset Confirmation packet. The effect of a Reset is to return the $P(R)$ and $P(S)$ counters to zero and to discard any data either in the network or stored in the DTE. The formats of the Reset Request and Reset Indication packets are identical except for the manipulation of the Channel numbers by the network as usual. The format of the Reset Indication/Request packet is now given:

Bits:	8	7	6	5	4	3	2	1	
Oct 1	0	0	0	1		LGN			GFI and LGN
2				LCN					LCN
3	0	0	0	1	1	0	1	1	
4				Cause octet					
5				Diagnostic octet					Optional in Reset Request

If the Cause field is set to zero this indicates that the Reset was instigated by the remote DTE, the DTE initiating the Reset is permitted to add one octet of user data referred to as the Diagnostic octet, this is passed to the remote DTE. If the cause octet is non-zero the Reset was initiated by the network, the value in the cause field defining the reason for the Reset, the Diagnostic field may in this case be used to provide network-specific information to add further information to the reason given in the cause field. The values for the Cause field for the PSS network are now given:

Cause octet	Resetting
8 7 6 5 4 3 2 1	cause
0 0 0 0 0 0 0 0	DTE initiated Reset
0 0 0 0 0 0 0 1	Out of order[†]
0 0 0 0 0 0 1 1	Remote procedure error
0 0 0 0 0 1 0 1	Local procedure error
0 0 0 0 0 1 1 1	Network congestion
0 0 0 0 1 0 0 1	Remote DTE operational[†]
0 0 0 0 1 1 1 1	Network operational[†]

† Indicates that these causes will only occur on Permanent Virtual Circuits where they are used to indicate the availability or not of the connection depending on the state of the X.25 link, this is necessary since PVCs do not use call set-up or clear-down in establishing a data path from DTE to DTE.

When a DTE receives a Reset Indication, whether it be generated by the network or by a remote DTE it should respond as soon as possible with a Reset Confirmation packet which informs the network that the DTE has reset its $P(R)$ and $P(S)$ counters and considers the call to be in a state identical to that of a call which has just been established.

The format of the Reset Confirmation packet is:

Bits:	8	7	6	5	4	3	2	1	
Oct 1	0	0	0	1	LGN				GFI and LGN
2				LCN					LCN
3	0	0	0	1	1	1	1	1	PTI

4.9 SUMMARY OF PACKET TYPE IDENTIFIERS

Packet type		Packet type identifier
		8 7 6 5 4 3 2 1
Incoming Call	Call Request	0 0 0 0 1 0 1 1
Call Connected	Call Accepted	0 0 0 0 1 1 1 1
Clear Indication	Clear Request	0 0 0 1 0 0 1 1
Clear Confirmation		0 0 0 1 0 1 1 1
Data		x x x x x x x 0
Receiver Ready		x x x 0 0 0 0 1
Receiver Not Ready		x x x 0 0 1 0 1
Reject		x x x 0 1 0 0 1
Interrupt Indication	Interrupt Request	0 0 1 0 0 0 1 1
Interrupt Confirmation		0 0 1 0 0 1 1 1
Reset Indication	Reset Request	0 0 0 1 1 0 1 1
Reset Confirmation		0 0 0 1 1 1 1 1
Restart Indication	Restart Request	1 1 1 1 1 0 1 1
Restart Confirmation		1 1 1 1 1 1 1 1

5

The Transport level

5.1 OVERVIEW OF THE TRANSPORT LEVEL

In this chapter we shall review the standards which have been defined and in some cases implemented which reflect the fourth layer in the ISO seven layer model. Strictly speaking the Transport layer sits not above the Packet layer we previously covered but above a Network layer. The network layer is intended to provide a common interface regardless of the underlying network technology which is in use. By defining a common interface to which both wide and local area networking protocols can converge, the whole spectrum of computer networks can be dealt with in one set of high level protocols. The work on the development of the Network layer is less well advanced than those above and below it. It will provide a virtual circuit capability as X.25 does, but with the ability to handle a wider addressing range, probably 32 digits rather than 12, and also longer expedited data messages of 32 octets rather than one. The problem area is how to map systems which are by their nature datagram networks rather then virtual circuit networks — Ethernets for example. When the deliberations on this are complete an ISO network will be able to span the range of networks from small intra-office local area networks to large-scale intercontinental wide area networks with one set of protocols and with the ability to interwork between them. This is a good example of the advantages of the layered architecture: the communications sub-layers and below can be replaced with another set of entirely different protocols structured in a different way to meet another set of constraints.

We shall cover in this section three interpretations of the Transport layer, one which predates the publication of the ISO Draft Proposal, created by Study Group Three of the Post Office PSS User Forum in the UK in 1980, the ISO Draft Proposal itself, and the European Computer Manufacturers Association (ECMA) Transport protocol ECMA-72, which is derived from the ISO work. The ECMA-72 standard will change (and has changed) to reflect the current thinking of the ISO committee TC97/SC16, the ECMA protocol omits some of the classes which the ISO propose and has a less rigorous procedure for the production of standards. They have essentially produced a subset of the final ISO specification for the Transport layer in advance of the ISO itself. The Transport protocol produced by the SG3 of the Post Office PSS User Forum is generally known as the Yellow Book (because of the colour of its cover), they have also produced a Blue Book File Transfer protocol, a Red Book Job Transfer and Manipulation Protocol, Green Book procedures for Character Terminals, and Grey Book Mail. To cover the Transport services defined by the ISO, ECMA, and the Yellow Book may seem a huge task, but since the ECMA protocols are a subset of the ISO definitions and the Yellow Book is rather simple the task will not, be excessive! In the final chapter (Chapter 6) we shall cover the 'Triple X' protocols defined for the connection of character mode terminals by a packet-switched network to a host computer system. The CCITT who created this protocol did not consider the use of any higher level protocols and simply defined their protocol directly above the X.25 packet layer. 'Triple X' then does not fit in well with the seven layer model view of communications; we include it here because it is in very widespread use and so is worth explanation. In the UK a version of the X.29 part of 'Triple X' to operate above the Yellow Book Transport Service (YBTS) termed TS.29 has been defined; the other two protocols of 'Triple X' remain unchanged; TS.29 provides the definition of how the characters are mapped into Transport Service Data Units. The rainbow book protocols we referred to above are in widespread use owing to the insistence of the Computer Board, who are responsible for the provision of computers in academic institutes and universities, that the machines supplied are equipped with the software and hardware to implement this set of protocols. As a consequence the UK has a very well integrated academic network in which users can sign on to remote computers interactively, or ship jobs, files or mail messages to any of the few hundred computers which comprise this network. These protocols have been adopted by the UK. Department of Trade and Industry as the basis for their 'Intercept' strategy to encourage the use of Open Systems Interconnection by the use of intermediate protocols where the final ISO protocols do not yet exist, gradually cutting over as the official protocols are ratified.

The *raison d'etre* for the Transport layer in the ISO OSI seven layer model is to provide some 'value added' functions above the functionality provided by the Network layer. Since the concept of datagram type network operation has been grafted onto the OSI work it is no longer assumed that the Network layer will provide a virtual circuit service to the higher layers and as a consequence the Transport layer protocols defined by the ISO and ECMA make allowance for the possible corruption of the Transport level data units by the underlying Network sublayers. The Yellow Book Transport Service makes the assumption that it will be operating on top of an X.25 network which provides it with an error free virtual channel and therefore does not include any error detection or correction code. The ISO and ECMA protocols for the Transport layer provide five different classes of Transport protocol to allow for a simple class of operation with no error detection correction, such as might be used for a Teletex service, through to a class offering error detection and correction, flow control, and multiplexing or splitting of Transport level connections onto the underlying Network levels calls. The purpose of the multiplexing classes is to try to optimise the use of the Network level according to either economic or throughput constraints. If the network operators charge a high rate for each virtual circuit in use but relatively little for the use of the connection then it may be financially advantageous to send several Transport level calls to the same destination all on the same Network level call. Alternatively, if an application requires high throughput or high reliability then it is also possible to split a single Transport level call over several Network level calls, possibly taking different routes to increase the resilience or capacity of the Transport level link. This is termed splitting. In outline the different classes of protocol for the ISO/ECMA transport service provide the following facilities:

Class 0   Almost a null protocol; only adds extra addressing capability and the segmentation of messages.

Class 1   Basic error recovery class; adds sequence numbers to all messages so that recovery from Network level resets or disconnects is possible using a Transport level Reject message to elicit re-transmission as in LAPB and X.25.

Class 2   Multiplexing class; does not provide error recovery but provides for several Transport level connections to be multiplexed onto a single network connection.

Class 3   Error recovery and multiplexing class, provides the functions of class 1 plus class 2.

Class 4   Error detection and recovery with multiplexing or splitting; the messages are protected by a checksum as well as a sequence number. The splitting could handle for example 50 transport level calls multiplexed down 10 Network level connections.

With these five different classes of operation it can be seen that the ISO/ECMA OSI protocols seem to introduce their own conformance and interworking problems — the higher classes are not simply supersets of the lower classes. The ISO view is that a system claiming conformance to the OSI protocols should implement at least one of Classes: 0, 01, 2, 23, 24. This means, however, that if one site supports only class 0 and another only classes 2, 23, or 24, they cannot interwork, rather defeating the object of the whole exercise! In the UK the Department of Trade and Industry is insisting that all implementations must support both classes 0 and 2, the others being optional in order to ensure that interworking is possible. Class 0 is designed to be compatible with the CCITT protocol S70 which defines the operation of Teletex terminals. The original S70 protocol caused the terminal to send its data directly over an HDLC link, the addition of a transport level 'wrapper' does not significantly increase the complexity of what are to be seen as relatively simple network devices. Classes 1 and 2 are suitable for operation over an X.25 packet-switched network in which error-free virtual circuits are provided, the multiplexing facility may be useful for cost mainimi-sation where there are several calls to the same network host. The addition of message sequence numbers which enable recovery from Network layer generated disconnects or resets is not expensive to implement. Class 3 may be used in such an environment if both error recovery and multiplexing are required. The class 4 protocol is not really appropriate to the X.25 network environment since it adds extra checksums and handles the re-sequencing of data. Checksums are typically quite expensive to implement unless hardware assistance is provided since every character in every message needs to be scanned, and out-of-sequence data should never occur without the Network level software issuing a reset. The checksum defined for use by the ISO/ECMA is the simple "exclusive ORing' of all the data octets.

This does not provide particularly good error detection compared with the cyclic redundancy checks operated by the LAPB protocol at the Link level. The intention of this protocol is to operate above an error-ridden non-virtual-circuit network such as the datagram facilities provided by the direct use of contention protocols for local area networks, Ethernet for example.

5.2 DESCRIPTION OF THE YELLOW BOOK TRANSPORT SERVICE

The definitive document describing the Yellow Book Transport Service (YBTS) is 'A Network Independent Transport Service' by Study Group Three of the Post Office PSS User Forum, prepared in February 1980. This document describes the facilities of the YBTS and also the mapping of these functions onto X.25. A multiplexing protocol and a protocol to implement YBTS above an asynchronous communications network are added as appendices which we shall not consider here as they are not yet widely used and are mainly included for discussion. The title of the definitive paper appears to suggest that the goal was to produce a protocol which is independent of the communications network upon which it is implemented; this is only partly acheived. The protocol is actually strongly orientated towards implementation over an X.25 packet-switched network, although it has also been implemented over Cambridge Ring local area network systems and in gateways between the two protocols. When implemented over Cambridge Ring Transport Service is known as Transport Service using Byte Stream Protocol (TSBSP). Byte Stream Protocol provides a virtual circuit protocol above the basically datagram-type network provided by the raw Cambridge Ring technology. In the UK many universities and other academic institues are interconnected by a network based on switching nodes made by GEC Computers Ltd and running software created both by the Science and Engineering Research Council (SERC) and GEC Computers Ltd. The nodes are interconnected by a variety of communications lines, ranging in speed from 2.4k bps to 48k bps and with considerable redundancy in connectivity. At the present time each institute is a separate address domain and therefore a gateway is needed between the local packet-switching exchange operated by the institute and the nearest JANET switching node. The gateway is a piece of software which resides in the packet-switching exchange and appears to both the inside and outside worlds as a host. X.25 does not handle the long addresses required in this situation (since the maximum address is 12 digits

plus a two-digit subaddress) so this problem is also resolved by the
YBTS. It is very rare that the address of the remote service being
called by a program making use of YBTS would consist solely of
the X.25 address, since most X.25 host machines will provide more
than a single function; and, even if they did, it would be nice to be
explicit what that function was. An example of this sort of simple
situation might be a remote printer on a network supported by a
dedicated microprocessor-based controller; it would still be useful
to identify the remote destination exactly in case another printer
is added at a later date, or if it were necessary to distinguish between
different character fonts being installed for example. In YBTS the
address is constructed from the route that the message will take,
in the order it will pass, and separated by a delimiter. The address
is to be composed of letters and digits where possible, the case of
letters is to be ignored and imbedded spaces are forbidden. The
delimiter may be one of '+/.;_,'. Thus, if a remote printer was on
an X.25 network as address 234290468168, the YBTS address for
it might be: 234290468168+LP1. Where a general-purpose system
is being called to perform File Transfer using the File Transfer
Protocol for which the recommended address is FTP the address
might be: 234210156221.FTP, if the file actually contained a mail
message to be forwarded by the mail system in the remote machine
the address may contain a hierarchy of routing to be followed once
the file is accepted by the remote File Transfer Protocol server,
e.g. 234272458158.FTP/MAIL. The Transport Service handler in
each machine that a Transport Service call passes through as it is
set up strips the leftmost part of the address up to the first delimiter
it finds and uses that address for the next hop of its route; this
applies whether the Transport Service is a server in a host machine
or a resident part of a packet-switching exchange. The Transport
Service may contain tables of names to translate from a title to an
actual address. The address for a call from a local machine out
through the gateway in the local packet-switching exchange to
JANET and then into a packet-switched network and finally to a
particular service in a remote machine might appear as follows:
JANETGATE+YORK+DEC10+FTP.

The Yellow Book Transport Service provides facilities so that as
an address is travelling through a network the address remains valid in
either direction. For example the address shown above might arrive
at YORK as YORK+DEC10+FTP, but YORK might actually have
two separate networks so that the gateway at YORK receiving the
address from JANET might have to insert another hop to get to the
right destination, when the address is returned to the originator

the complete route may be spelled out so that if the network had failed the call could be re-established. This is particularly helpful if the protocol running over YBTS makes use of checkpointing so that in the event of a network failure the interaction can be restarted from some known point. The called address could request some generic service such as FTP which is mapped by the Transport Service onto some specific invocation of the server such as FTP_3. YBTS provides the following primitives for high level protocols to make use of:

CONNECT	To request a connection to a remote service.
ACCEPT	To accept an incoming CONNECT.
DISCONNECT	To reject a CONNECT or clear an existing connection.
ADDRESS	To pass an address to be used in a CONNECT.
DATA	To send data to the remote service.
PUSH	To force delivery of any data in transit.
EXPEDITED	To send data outside normal flow control constraints.
RESET	To clear down a connection to a known state.

The definition of YBTS does not specify how these functions should be provided to high protocol servers within a host machine. For some applications a simple synchronous 'CALL' type of interface may suffice but for multi-threaded applications an asynchonous type of operation is required with some sort of flow control so that the task can control the rate at which data is presented to it. This is typically achieved by having the YBTS user task present it with buffers to be filled with data and to raise a software interrupt or set a semaphore when the buffer has been filled. YBTS does not consider data to be divided into records, but just as a continuous stream. If data is presented for transmission by the YBTS in a host by a series of DATA messages of a small size the system may buffer them and send them through the network as a single large packet, or vice versa; it is up to the high level protocol to impose some sort of structure by inserting markers into the data. The PUSH function is provided to allow a program to elicit a response from another by forcing any data which is being buffered by the host YBTS or any intermediate nodes to be forwarded for delivery to the remote task. Since this function causes the transmission of partly filled packets at the network layer it is inefficient to use this to provide a record delimiter for higher level protocols. This is an example of the orientation of this protocol to the X.25 environment, we shall see the reason for this dependence when we study the mapping of the YBTS primitives onto X.25

shortly. EXPEDITED type data is needed since we have indicated the necessity for flow control, if we need to signal the remote system but it is unable to accept more data we should be stuck if it were not for this ability to send data regardless of flow control constraints. The X.25 Interrupt facility allows for the transmission of one octet of data outside flow control and hence that limit also applies in YBTS which is so heavily dependent on X.25. For the same reason only one EXPEDITED message can be outstanding in each direction of a connection at any time. The YBTS RESET function is mapped on to the X.25 Reset function and is used to resynchronise the two ends of a conversation after some catastrophic failure. RESET can either be generated by a higher level protocol making use of YBTS after it has recognised that the two ends of the conversation are out of step in some unrecoverable way, or by the YBTS service itself if it detects that the Network layer beneath it has reported that data has been lost in some irretrievable way. The effect of a RESET is to clear down any data which was in transit or waiting in buffers at any intermediate node and reset the link so that it is in the same state as when it was first created by the exchange of a CONNECT and ACCEPT message pair.

5.3 REALISATION OF YBTS UPON X.25

In general Transport Service data units and control messages are sent using Network Service data units; however, the CONNECT message may be carried as part of the level 3 Call Request packet as we shall describe a little later. We shall assume initially that the level 3 call has been established. The Transport Service messages are divided between the data-carrying message types, i.e. DATA and PUSH, and the control message types, i.e. CONNECT, ACCEPT, DISCONNECT, RESET, ADDRESS and EXPEDITED. DATA and PUSH are mapped directly onto the level 3 data stream without the addition of any framing data and is consequently very efficient in its use of the level 3 facilities provided. Normal DATA messages are sent as Data packets on the X.25 Network layer, they are sent with the Qualified data bit set OFF and the More data bit set ON.

Since the More bit is set ON the X.25 Packet level software is forced to buffer enough data to send only full packets out into the network, this being mandatory if the More bit is on. The two ends of the conversation may be working with different packet sizes, the network connecting them resolving the differences by an attempt to send messages of a particular size. To ensure the delivery of data by the YBTS server the PUSH message is sent whenever a reply to the data sent is required. The PUSH message is encoded by causing any data buffered by the Network layer to be sent as a data packet with the More bit set OFF. All intermediate nodes will, on receipt of a data packet with the More bit set OFF, immediately forward any data they are buffering for that conversation. Since the PUSH is delivered at a particular point in the data stream to the remote end of a connection, it can be used in applications where some sort of record delimiter is required; however, because partly used packets are sent through the network as a result, it is inefficient to use it in this way. Generally a higher level protocol will make use of its own markers in the data stream and only use the PUSH function when a reply is needed to some message such as at the very end of a file transfer to ask if the data has been filed away correctly and it is all right to mark the transfer as having been completed. The control type messages make use of a length and flag octet at the start of each parameter fragment which is termed a 'header' octet. Each parameter can in theory be of infinite length; in practice a limit of 255 octets has been agreed as a practical limit. The header octet is made up of the following eight bits: /E C N N N N N N/ where:

E − Indicates the end of the parameter if 1.
C − Normally zero but used in the CONNECT message type.
N − Six-bit length field for this fragment of the parameter.

The control messages are always sent as Qualified data messages, if the parameters are of such a length that more than one packet will be needed to send them, then the first and any intermediate packets will be sent with the More bit set ON as well as the Qualified data bit, the last or only packet of a sequence will have the More data bit set OFF. The first octet of a control message is always a code which indicates which of the control messages is being sent. The values of the codes are as follows:

CONNECT   16
ACCEPT   17
DISCONNECT   18
RESET   19
ADDRESS   20

A null parameter to a control message is indicated by sending a header octet with the end of parameter bit set ON and the length field set to zero; the length field does not include the header octet itself. If the parameter is trailing and null then it need not be sent at all. If the parameter is longer than 63 octets, then it is divided into fragments each with a maximum length of 63 octets, and each with its own header octet in which the 'E' bit is set to zero in all but the last fragment of the parameter. In this way parameters of indefinite length may be transmitted. In practice, however, a maximum length of 255 octets per fragment is adhered to by the user community so that enormous buffers do not have to be reserved for them. If a CONNECT-type message is being sent by a system which makes use of the Fast Select facility at the Network level, the parameters may make use of the 124 octets available after the four octet protocol identifier, which is set to 7FFFFFFF to identify the call as a Transport Service type call. If fast select is not being used, the remaining 12 octets may be used to carry the first fragment of the CONNECT message.

5.4 STRUCTURE OF YBTS CONTROL MESSAGES

Having indicated the way in which the different YBTS control messages and their parameters are transported by an X.25 Network layer, we now describe in detail the information which is passed from end to end by these control messages. Although YBTS was designed with the facilities provided by X.25 in mind as a constraint on its facilities, it is by no means the only environment in which it has been implemented; it is also to be found on Cambridge Ring local area networks, and also in connecting intelligent work stations by means of asynchronous ports on Packet Assembler/Dissassemblers.

The different message types are described with their parameters in the order in which they are expected in the following format:

    COMMAND (Parameter 1, Parameter 2, . . .)

where COMMAND represents the name of the YBTS control message, and Parameter 1 etc. represents the parameters in the order in which they must be sent. Only parameters which are underlined are obligatory.

CONNECT (Called Address, Calling Address, Quality of Service, Text)

The most important parameter of the CONNECT function is the Called Address which indicates the destination of the proposed connection. The transport service address at this point consists simply of text as inserted in the Connect request made by the higher level protocol making use of the transport service. The function of the Transport Service in the machine which generated the request will be to take the left-most part of the address, up to the first

delimiter and use it to make a Network level call to the next hop
en route to the required destination. The address might be for
example:

234241260106+71000013+FTP+MAIL.

In this case the YBTS program would use the address 234241260106
to make an X.25 call passing the string: 71000013+FTP+MAIL as
the Called Address to the next stage in the route. The software
receiving this incoming call from the network would then itself make
an outgoing call to the address 71000013 with a Called Address
parameter of FTP+ MAIL and so on. In the same way the
Calling Address starts as either null or simply the identification
of the task making the call request and is gradually appended to as
the Connect passes through the various networks on its way to its
final destination, so that on reaching it the Calling Address field is
a complete route map of the passage of the call through the network.
The Quality of Service parameter is not used on X.25 networks at
present since there is no standard way of requesting that different
calls be treated differently from each other regarding throughput or
transit delay etc. If this facility is ever introduced the text would
contain a code to be passed to the Network layer to indicate which
class of service this call should make use of. The Text field can be
used to pass a string of data from caller to called for information
purposes, but should not be used as part of a high level protocol
since the delivery of the string is not guaranteed.

ACCEPT (Recall Address, Quality of Service, Text)

The normal response to a CONNECT is an ACCEPT message. None
of the parameters is obligatory; however, the Recall Address is
frequently useful and is generally sent. Typically a high level protocol,
such as for file transfer or interactive access to a host computer, will
make a Connect request to a generic service identifier rather than
a particular invocation of that service; for example it might call:

71000013.FTP

to access the remote file transfer server task. If this is a multi-threaded
task it may be dealing with file transfers to and from several other
sites at the same time and so might return a Recall Address of FTP_5,
the passage of this message back through the network will cause it
to be translated in the same way that the Recall Address on the
Connect was built up as it passed through the network, so that it
might be returned as 71000013.FTP_5. If the network connection

were to fail during the duration of a long data transfer, the high level protocol handler for the file transfer facility could re-establish the Network level call, this time quoting the exact destination address rather than simply the generic address. Thus the transfer could be restarted with the correct thread of a multi-threaded file transfer server by resuming at some checkpoint agreed by the two protocol handlers. The Quality of Service is returned in case the passage through the network had caused the value to be negotiated to some reduced value. The Text parameter is again used to pass information from one end of the connection to the other, for example to give a textural confirmation of the identity of the remote system which has been accessed.

DISCONNECT (Error Code, Address, Text)

DISCONNECT is used either to reject an incoming Connect request or kill an established call. The obligatory Error Code is a single octet defined in the reference document as having various positive values. Zero implies a successful end to a call; one is the response to an incoming Disconnect; other positive values indicate either the reason for rejection of a Connect request or the cause of an unsolicitied Disconnect. In minimal implementations the Disconnect function may be mapped directly onto a network level call clear, passing the error code in the diagnostic field of the Clear Request packet. If the full protocol is implemented the Address parameter may be used to indicate where several networks are being traversed and a node or gateway in one is going down since it can be identified by the network address which is given in this parameter. The Text field as usual can be used to pass information of a non-critical nature from end to end of a connection. When a Disconnect control message is sent, as opposed to a level 3 clear request, a timer is set so that if flow control constraints prevent the delivery of the disconnect message a Reset can be sent which will cause any data in transit to be thrown away, the code sent on the reset will indicate that the connection is now 'Out of Order'. If the reset request fails to elicit a disconnect response from the remote system the level 3 Clear Request may be issued as in the case of the minimal implementation recommendations.

RESET (Error Code, Address, Text)

The RESET function is used to clear down a connection either when it has become stuck in some way or where the two co-operating high level protocol modules at each end of the link have become so out of step that the only means of recovery is to throw away any data which is in transit or buffered anywhere in the network and restart themselves from some checkpoint. After a Reset has occurred the connection is in the same state as it would have been if it had been cleared and then re-established. The Error Code parameter is a single-octet code indicating the cause of the Reset; the Address parameter gives the textual address of the originator of the Reset — this could be either end of the connection or one of any intervening transport service gateways or packet switches. The text parameter as usual may be used to carry data explaining the cause of the Reset in human readable form. It should be noted that the textual information parameters carried by the control messages are always optional and, even if the Transport service handlers at both ends of a link can carry the textual message, it is possible that the nodes *en route* may not, hence the textual information field may not be used by any high level protocol to carry data essential to the operation of the protocol if complete connectivity is being aimed for. The transmission of a Reset at the Transport level is accompanied by a Packet level reset which actually performs the clearing down of any buffered data in transit. In a minimal implementation the Transport level control message may be completely omitted and only the Packet level reset sent, using the diagnostic field to carry the machine-orientated error code.

ADDRESS (Address, Qualifier)

The ADDRESS primitive is used to exchange addressing information; as the addresses are passed they are subject to the usual appending of each part of the address as they pass through a multi-hop network. The Qualifier is used to indicate whether the address is passing towards or away from the addressed object, a value of 0 indicating that the address is moving towards the addresses object. 1 that it is moving towards the addressed object, and 2 if an addressing error has been detected. This method of passing addresses of objects around the network has never been implemented and in only included here for completeness.

EXPEDITED (Data)

The EXPEDITED function is used to pass information from one end of a connection to the other while over-riding any flow control hold-ups that may be in progress. A typical use would be to interrupt a program running on a host computer from a network-connected terminal. If a program has been initiated on the host it may not put

out any read operations which would cause Transport level data
units to be read in from the YBTS handler, in order to 'Break in'
a special message needs to be sent which can get through even if
there is no read outstanding for that terminal, or if flow control on
the connection is preventing any further data from being input.
The Expedited function is used for this situation where an immediate
response is required in real time, rather than at some point in the
data stream. The function is unlike the preceeding ones in that it
is not mapped on to a qualified data packet at the packet level but
is instead mapped on to the X.25 Interrupt facility. For this reason
only a single data octet can be transmitted on an Expedited data
message, and only one can be active in each direction at any time
If a larger message is passed to the transport service handler then
it is sent one octet at a time subject to the above-mentioned constraint. Typically no data at all is carried by the Expedited message,
but the receipt of the message type is sufficient to cause the Break-in
action required.

5.5 DESCRIPTION OF THE ISO TRANSPORT SERVICE
The level 4 Transport Service protocol defined by the International
Standards Organisation is defined by two documents produced by
Sub-committee 16 of Technical Committee 97 (often referred to as
TC97/SC16) which define the Transport Service Protocol Specification and the Transport Service Definition. The former describes
the implementation of the Transport Service above a Network
layer based on X.25 but enhanced to provide extra function in
accordance with the definition of the Network layer, the latter defines
the service provided by the Transport Service. In this description
we shall merge the discussion of protocol operation with descriptions
of the function of the various primitive operations provided by the
Transport Service, in order to describe the reasons behind the operation
of the protocol in terms of the problems which are solved by its use.
The Transport Service is mainly concerned with the provisions of
extra quality of service over that provided by the raw network service.
In particular the Transport Service is intended to make it possible to
optimise the use of the network layer with regards to its tariff structure, the requirement of different higher level applications for different
throughputs error rates, reliability or data security. In common
with the previously described Yellow Book Transport Service the
protocol provides a transparent data stream from end to end with
the ability to split up messages by the use of a logical end-of-message
marker. The basic unit of data transmission at the Transport layer is
the Transport Service Data Unit (TSDU) which must consist of

an integral number of octets of user data. The quality-enhancing activities of the Transport Service can be divided into those occurring at Transport connection establishment, during the transmission of Transport Service Data Units, and during the termination phase of the connection. The level of improvement in the quality of service provided depends on the particular class of Transport protocol which has been requested from the five that are defined. The five classes are referred to as classes 0–4; they are not all supersets of each other as the class number rises; these discontinuities mean that there are possible incompatibilities if different machines implement different incompatible classes of the Transport Service protocols. The classes are defined as:

Class 0 – Simple class
Class 1 – Basic error recovery class
Class 2 – Multiplexing class
Class 3 – Error recovery and multiplexing class
Class 4 – Error detection and recovery class

In all of these classes there is a skeleton of basic functions which are common to all of them. These are the assignment of a Network layer connection to carry the connection, the establishment of Transport level connections, the possible refusal of incoming Transport connection requests, the release of connections, the transmission of data and its segmentation if the data unit is larger than the underlying Network data unit, and the handling in some way of protocol errors. We shall now tabulate the differences between the sub-functions provided by the various classes of Transport connection and describe the function and operation of these sub-functions (see Table 5.1).

The function and implementation of these different sub-functions will now be discussed and the resulting concepts synthesised into a description of the operation of each of the five classes of protocol. The term Transport Protocol Data Units (TDPU) is used in the following section to refer to the complete Transport Service message, not merely the user data it may be carrying. TPDUs are divided into two types: Data TDPUs, and Control TDPUs. Control TDPUs may contain a User Data field whose length is indicated by the length of the Network Service Data Unit which is being used to carry the message; the different control messages are either of fixed length or have a length count and does not have a explicit length but is defined by the length the Network level message in which it is carried.

We first describe the protocol mechanisms which are provided in all the different classes followed by those which are implemented in some classes only.

Table 5.1.

Sub-function	Class	0	1	2	3	4
Concatenation and separation			Y	Y	Y	Y
Implicit termination		Y		Y		
Numbering of Data messages (Normal)			Y	M	M	M
(Extended)				O	O	O
Expedited Data using N-Expedited				N		
not using N-Expedited			M	Y	Y	Y
Reassignment			Y		Y	
Reassignment after failure			Y		Y	
Holding of TSPUs until acknowledgement						
(Conf. Receipt)			N			
(using AK)			M		Y	Y
Resynchronisation			Y		Y	
Multiplexing				Y	Y	Y
Use of Explicit Flow Control				M	Y	Y
Use of Checksum						M
Non-use of Checksum		Y	Y	Y	Y	Y
Frozen References						Y
Retransmission on Time-out						Y
Resquencing						Y
Inactivity Control						Y
Splitting						Y

Y = sub-function is included in this class
M = mandatory sub-function, use is negotiable
O = optional sub-function, use negotiable when implemented
N = optional sub-function dependent on Network layer facility, use
     is negotiable when implemented

Network connection assignment

Before a Transport level conversation can take place the Transport Service must have assigned the conversation to one of the available Network level connections. It may either choose to use an existing connection, if one exists to the required destination and multiplexing has been specified as an option when the protocol class for the Transport connection has been chosen. Several Network level connections may be set up if splitting has been defined as a requirement for the Transport level connection; this will cause several different

Network level connections to be set up to the same destination in order to increase the throughput or reliability of the service obtained. An existing connection to the required destination may be rejected if it does not have the 'Quality of Service' requirements associated with the required Transport level connection. When all the Transport level connections to a particular destination have been terminated the Transport service may choose to keep the Network level connection open for a period of time in case a further Transport level connection to that destination is requested.

TPDU Transfer

The Network level Data and Expedited Data functions are used to convey Transport Protocol Data Units across the established network connection. The possible TPDUs which can be transferred are:

Connection Request	CR
Connection Confirm	CC
Disconnect Request	DR
Disconnect Confirm	DC
Data	DT
Expedited Data	ED
Data Acknowledge	AK
Expedited Acknowledge	EA
Reject	RJ
TPDU Error	ERR

Network level Data messages are used to convey all of these TPDU types except for Expedited Data and Expedited Acknowledge in the case of class 1 where the use of Network level Expedited may be negotiated.

Data TPDU segmenting

The User Data field of a Data TPDU may be of any length up to that agreed as a maximum when the connection was established. The Data TPDU contains an End of Transport Service Data Units mark whose function is almost analogous to that of the More data bit in the X.25 packet interface. A message may be split into several Data TPDUs each with the End mark set to NO except for the final one which will be set to YES to indicate that the message is complete. Unlike the X.25 equivalent, the Data TPDUs that precede the final one do not have to be filled. If the message will fit into one Data TPDU then it will have the End mark set to YES.

Concatenation and separation

Where it is possible, several Transport Protocol Data Units may be sent in a single Network level packet. Each TPDU has a length indicator in its header which accounts for the Transport Service protocol messages but not for any user data, which is delimited by the length of the underlying Network Service Data Unit. If several TPDUs are to be concatenated into one NSDU they must either all be control type TPDUs with no User Data field, or, if there is user data to be carried, it must be in the last position of the NSDU. This is so that the length of the user data can be determined by the length of the NSDU, if other control messages were allowed to follow user data it would not be possible to determine where the next control message after the data started. Where several TPDUs are concatenated the overall length will not exceed twice the agreed maximum length of a Transport Protocol Data Unit agreed.

Transport connection establishment

The connection Request is passed in a Network level data message, the creation of Network level connections is a separate occurrence not related to the establishement of Transport level connections, which may or may not cause a new Network level connection to be established. When a Connection Request (CR) is sent it may contain several parameters, most of which are optional; we shall start by describing those which are essential and then cover the optional ones. Each Transport service chooses a Reference which is sent to the other end of a connection, i.e. the initiator of a Transport connection sends a Reference on the CR and the acceptor of the request returns another in its Connection Confirmation (CC). The reference is a sixteen-bit number greater than zero and not already in use or classed as frozen. Since each partner in the connection informs the other of the identification it expects when any message concerning the particular Transport connection is transferred, there can be no problems of call collision such as occur with the fixed scheme used by the X.25 packet level. The calling and called addresses of the required Transport Service Access Points are exchanged and are obviously used by the Transport Service to determine which Network level connection to use or whether to establish a new one. In classes 2, 3, and 4, where explicit flow control may be used, the initial credit of tokens permitting messages to be sent may be exchanged when the call is established. In all classes except class 0, up to 32 octets of user data may be exchanged using the CR and CC functions. The initiator of a Transport connection requests a particular protocol class which it would prefer to use for the connection, and possibly some alternatives, except that, if the request is for class 0, no alternatives can be given. The responding Transport Service may choose either the requested class of protocol or one of the suggested alternatives; or, if classes 3 or 4 were proposed, it may request that class 2 be used; or, if class 1 is proposed, that class 0 be used. In the United Kingdom it is required for conformity that Transport services should implement both class 0 and class 2 in order that full interworking may be possible. The maximum size of TPDU that will be permissible is from the range 128 to 8192 octets, and powers of two in between; the responding system may reply with a proposed maximum TPDU size anywhere between the size proposed and 128. The length of the sequence number field may be either 7 bits or 31 bits in length, which are termed normal and extended respectively; if extended sequence counts are requested then the initial credit of tokens for explicit flow control will be an extended field also. A checksum selection parameter expresses the desire to use checksum protection on all Transport level messages carried on this Transport connection. A version number may be sent on the Connection Response so that the initiating Transport Service may check that the remote system is using a version of the protocol which is compatible. A security parameter is provided which has a user-defined syntax in order to allow the higher level protocols to pass information about any encryption system they may be operating. The Quality of Service which a particular Transport Service connection requires is defined in a Quality of Service parameter which defines the throughput, delay, priority and residual error rate which are acceptable to the user of this connection. The remaining parameters define exactly how the Transport Service makes use of the underlying Network services, for example to negotiate the use of Network level Expedited Data to carry the Transport level Expedited type message. Other Network level facilities, which may be used by negotiation, provide for the use of the Delivery bit option in the Network layer to provide confirmation of delivery rather than the use of explicit acknowledgement messages at the Transport level, and the use or non-use of explicit flow control in the Multiplexing class 2 protocol. The non-use of explicit flow control in the multiplexing version of the Transport Service implies that the only way to hold back data flows is to apply flow control at the Network level; this has the side-effect of reducing or stopping flow on all the Transport level connections making use of that Network connection.

Connection refusal

If the Transport Service receiving a Connection Request is unable to accept the call it may reply with either a Disconnection Request or an Error function. In either case a reason code is passed together with up to 64 octets of user data in the former case while in the latter case the parameter which the Connection Request contained which appeared to the recipient to be in error is returned. When the initiating Transport Service receives either of these responses to a Connection Request the connection is regarded as closed and the source reference is available for re-use. The reason code used is a single-octet, machine-oriented error code.

Connection release

When a higher level protocol wishes to terminate a Transport level connection it does so by using the Release function. The Disconnection can be achieved in two ways: either by the disconnection of the Network level connection which is carrying the Transport level connection, or by the exchange of a Disconnect Request and Disconnect Confirmation message pair at the Transport level. A single-octet reason code may be passed and, if the Network level Disconnection is not being used, up to 64 octets of user data may be passed. The reason for permitting the passing of user data on Disconnects, including those in response to Connection Requests, is to permit a short transaction to take place without the overhead of having to completely establish a Transport connection. A short transaction might be initiated for example by a credit card validation terminal which will send to some central system the details which can be read from a magnetically encoded credit card and receive in reply a confirmation that the amount of credit requested by the customer falls within their credit limit at that time. This sort of transaction requires only a small amount of data and only a single record to be transmitted in each direction on the link.

Implicit termination

This function causes the Transport level connection to be terminated whenever the Network layer signals either a Disconnect or Reset indication. When either of these messages is received at the Network level the Transport level connection is terminated and the reference is free to be re-used. In practice, however, the use of the reference should be postponed for some period of time since the remote system may take some time before it realises that the connection has failed.

Spurious disconnect

It is possible that a Disconnect Request may be received by the Transport layer for a connection which does not exist. If the spurious disconnect occurs in a multiplexing protocol class then a Disconnect Confirmation using the same reference as in the Request should be issued. In any other conditions this situation should be treated as a protocol error.

Data TDPU numbering

The numbering of Data TDPUs by the Transport service can be used to provide flow control, to resequence messages which have been split over several network connections which have slightly different throughput characteristics causing some messages to overtake others, and in recovery from data loss or network connection failure. The numbering can be either normal or extended depending on the number of bits used to carry the sequence number. Normal sequence numbering makes use of modulo 2^7 arithmetic, i.e. the values cycle from 0 to 127 and then back to zero, whereas Extended numbering uses modulo 2^{31} arithmetic.

Expedited Data Transfer

The Expedited Data Transfer function provides a mechanism for sending data to the other end of a link apart from the normal flow of data. If, for example, the flow of normal data is blocked by flow control, Expedited Data will still be able to get through. At all points in the network Expedited Data is given priority over normal data transfer. The data may be transferred using the Network level data function to carry it, or by the use of the Network level data function to carry it, or by the use of the Network level Expedited facility where this exists and has been negotiated between the two ends of the link. Each Expedited Data unit which is received must be acknowledged by the return of an Expedited Acknowledge function to the originator of the message.

Reassignment

If the Network level connection which a Transport level connection was making use of is to be replaced by another, for example, if the Network connection carrying the Transport connection has failed, then a new Network connection may be established to carry the connection. On the new connection a Transport level Connection Request is issued with the Reassignment parameter showing the Reference value of the Transport connection which has been reassigned to this Network level connection; the receiver of this special Connection Request, which is really reassigning an existing Transport connection to a new Network connection, can update its tables to reflect the new Reference number which is related to the connection. Having established a reassignment the Transport service will then cause a resynchronisation to occur, in order to recover any data which may have been lost when the old connection failed.

Reassignment after failure

When a Network level Disconnect occurs on a channel being used to carry a Transport level connection, the initiator of the connection must reassign it to a new Network level connection. If the Reassignment is not established after a mutually agreed period of time, then both ends of the connection will consider the connection to be dead.

Retention of TPDUs until acknowledgement

In order to provide for recovery from error situations where data is lost by the Network layer, the Transport level may keep copies of certain TPDU types until some acknowledgement is received for them, at which point they can be discarded. The TPDUs to be retained are. Connection Request, Connection Confirm, Disconnect Request, Data, and Expedited Data. The acknowledgement may consist of the receipt of a Transport level message confirming the message has been received, or by means of a Network level delivery confirmation facility. In X.25 the facility is provided by the reception of a packet in the opposite direction with the $N(R)$ field showing a value greater than that of the $N(S)$ in the packet which was sent.

Resynchronisation

Resynchronisation occurs after an event in the Network layer makes it likely that data has been lost, after a Network level Reset has been received or a Transport service reassignment has happened. The Transport service which has initiated the Resynchronisation, for example, the one which sent the Reassignment, will retransmit a Connection Request or a Disconnect Request if it has an unacknowledged one outstanding, otherwise data transmission is resynchonised by sending a Reject TDPU with the sequence number field set to the next expected sequence number. Any Data or Expedited messages which are held by the Transport service are retransmitted. Any Data or Expedited messages received with a duplicate sequence number are discarded, except that any Expedited Data messages should first be acknowledged.

Multiplexing and demultiplexing

This function permits several Transport level connections to make use of the same Network level connection. No special parameters are required for this function since all Transport message types carry a reference number which identifies the Transport level conversation to which the particular message refers.

Explicit flow control

In order to control the flow of data from one Transport entity to another, class-dependent mechanisms exist to regulate it. This flow control is independent of the flow control in lower layers. If there is no multiplexing of Transport calls over Network connections then there is nothing to be gained over simply blocking the flow of data at the lower level; but if multiplexing is taking place, this would cause all the connections being carried by the lower level to be blocked which is undesirable and could lead to a deadlock situation. Class 0, which is not a multiplexed protocol, makes use of Network level flow control, as does class 1. In class 2 the use of explicit flow control is optional; use of the Network level flow control mechanisms is permitted. Classes 3 and 4 always make use of explicit flow control. When explicit flow control is in use the receiver maintains a variable containing the next expected Received TSDU sequence count, termed $T(R)$. The receiver indicates to the remote end of the connection the number of TSDUs that it has buffer space free by sending a credit allocation using the credit field of the Acknowledge TPDU. The size of the permitted window is determined at Connection Request time by the size of the initial credit field. The same windowing mechanism as is used in levels 2 and 3 are applied here using the 7- or 31-bit sequence counters and the credit field of the AK message.

Checksum

The checksum option is used where the Network layer carrying the Transport level connection may be unreliable; it is a variant on the longitudinal checksum. The entire TPDU is assembled and the value zero is placed into the two octets which form the checksum, then for the entire length of the TPDU two variables termed $C0$ and $C1$ are accumulated. The two variables are initialised to zero; the value of each octet in the TPDU is then added to $C0$, which is in turn added to $C1$, both using modulo 255 arithmetic. The two octets of the checksum field are then computed; they are termed X and Y for the first and second octet respectively:

$$X = (((L-n) * C0) - C1) \bmod 255$$
$$Y = (((L-n+1) * (-C0)) + C1) \bmod 255$$

where n is the offset in the TPDU of the first octet of the checksum and L is the length of the TPDU.

If this algorithm cannot produce the correct result for X and Y then the TPDU must be discarded. If the TPDU was multiplexed with any others in the Network level message they must all be discarded since it is not possible to trust the reference value since it may have been corrupted. This is treated as though a Network level error has occurred and resynchronisation has been attempted.

Frozen references

When a Transport level connection is released, the reference associated with it may not be re-used until some period of time has elapsed in order to ensure that the remote system has had time to determine that the connection has been terminated. In the case where the termination is caused by the clearing of the Network level call the reference is not frozen since the network will inform the remote system by passing it a Network level Disconnect Indication.

Retransmission on time-out

When a Transport service has outstanding Data or Expedited Data TPDUs which have been retained awaiting acknowledgement for greater than a timer termed T1, they will be retransmitted (up to the limit imposed by the windowing system) a maximum of N times, at which point the connection is determined to have failed and the connection release phase is entered.

Resequencing

The receiver in a Transport connection has the responsibility for ensuring that TPDUs are passed to the higher levels in the communications system in the correct order. The TPDUs contain a sequence number to facilitate this. Typically this problem should not occur where a good quality Network service is being employed, except in the case of the class 4 protocol where splitting of one Transport service connection over many Network level connections is taking place. Under these circumstances it is quite possible for mis-ordering to occur if the different Network connections pass through the underlying network by different routes.

Inactivity control

In the Class 4 protocol a mechanism is provided to detect a failure of the Network level which is not reported to the Transport layer. This is implemented by agreeing on a timer whose expiry indicates that the Network level link has failed. In order to prevent the expiry of this timer when there is no user data being sent over the Transport service an AK must be sent at intervals often enough to ensure that the inactivity timer never expires.

Handling of protocol errors

Whenever the Transport service receives a TPDU which contains a protocol error of any kind, such as invalid format or sequence number error, it will return an Error (ERR) function. The ERR function passes a cause code and the TPDU which was received in error up to and including the erroneous octet. As an alternative the connection may be simply released by sending a Disconnect Request in a minimal implementation.

Splitting and recombination

When class 4 has been selected, the splitting option may be used to provide greater throughput and reliability for a Transport connection. In this function the Transport connection is allocated to several Network level conenctions. TPDUs for the connection may be transmitted on any of the available Network connections. Since different TPDUs may travel through the network by different routes with differing delay characteristics the TPDUs may arrive at the destination address in an order different from that in which they were sent, the resequencing function will cause them to be reassembled in the correct order before they are passed to the user of the Transport service. Messages are never passed to the user from the Transport service out of sequence.

5.6 DESCRIPTION OF THE ECMA TRANSPORT SERVICE

ECMA is an acronym for the European Computer Manufacturers Association, which is defined a Transport Service standard termed ECMA-72. For those fearing that this chapter will interminable it will come as good news to hear that the ECMA-72 Transport Service standard is a subset of the ISO Transport protocol which we have just described. The ISO work has been adopted at various stages in its development cycle as a working standard on which products may be built. Although the ISO Transport Service has not been through its final stage in being published, at the time of writing, the ECMA standard took very early draft stages in the development as the basis of its work. As a consequence the standard has been revised to reflect the current level of the ISO Transport Service. The major difference with the current standard ECMA Transport Service is the lack of any class 1 protocol. In all other respects the document describes the same system as those of the ISO but in rather terse terms.

6

'Triple X'

6.1 INTRODUCTION TO 'TRIPLE X'

'Triple X' is the colloquial name given to the three protocols defined by the CCITT to connect character mode terminals to packet-switched networks using the X.25 packet definition. These protocols are:

X.3   defines the service provided to the terminal by the PAD
X.28 defines the user interface to the X.3 services
X.29 defines the usage of X.25 packets to carry the data.

The PAD mentioned above is the device which matches the characteristics of the two communicating services, the dumb character mode terminal on one hand, with the complex packet-switched network on the other. The term PAD is an acronym for Packet Assembly/Disassembly. Packet-switched networks are inherently block mode systems which operate most efficiently when handling reasonably large pieces of data. A character mode terminal is typically equipped only to receive and send one character at a time. The PAD is a piece of software and hardware which spans this gap in intelligence by providing a buffering service between the character-oriented terminal and the block-oriented packet-switched network. When character mode terminals are used in a simple environment they are connected directly to a computer system by a multiplexer; the computer system receives each character as it is typed. Generally some character is chosen to represent the end of a message so that each character is buffered up until this terminating character is received, at which point the buffer is passed to the software owning the terminal to be acted on. In the reverse direction the computer software will present to a system service a complete buffer of data to be sent to a terminal which is then transmitted to the terminal one character at a time. A simple system as described could be easily implemented in a packet-switched network by having a simple PAD which would buffer characters received from the terminal until the defined terminating character is received, at which time the buffer would be forwarded as a packet through the network. However, software is rarely, if ever, that simple; different software on the same computer system may have different conventions about what character will terminate messages, let alone the widely differing conventions that can be found between different manufacturers' systems. The 'Triple X' protocols are an attempt to provide a general-purpose solution and, as such, inevitably have to provide a compromise at times between usability and generality. Since this protocol is defined by the CCITT and not the ISO it does not form a part of the 'seven layer model, nor does it easily fit in with its structure. It is described here since it is a very widely used standard over X.25 networks and it would be churlish to exclude it on the grounds of 'purity'! The present 'Triple X' protocols specify that messages to and from the PAD are to be directly encoded onto X.25 packets, with no mention of a Transport Service to carry them. In the United Kingdom the X.29 protocol has been used above the Yellow Book Transport service successfully by the mapping known as TS.29 in which Transport Service Data Units rather than X.25 packets are used to carry the messages. The X.29 protocol only makes use of the Qualified and More data bits in the packet header, in TS.29 the More data bit is irrelevant since a TSDU can be as long as necessary and the operation of the Qualified data bit is simulated by the use of an octet at the start of each TSDU, which is either zero for Unqualified data, else 128.

We shall proceed in this chapter to describe the process of setting up and clearing down a virtual call on behalf of a terminal and then show how the session with the remote host is customised to make the PAD look as much like a direct connection to the host as possible.

6.2 ESTABLISHING A CONNECTION TO A REMOTE HOST

As soon as we start to describe the sequence of events as driven by commands at a user's terminal we shall see that an international committee is not necessarily the best group of people to describe user-friendly software interfaces. The use of the software interface to be described is not mandatory and most PADs which are provided by commercial manufacturers for private networks are much more user-friendly; those provided by PTTs as part of a national packet-switching service generally make use of this unpleasant interface, however. In order to identify yourself to a PAD you must be a registered user (again this does not always apply to private PADs) so that you can be billed for resources used, this identification is known as an NUI or Network User Identification. The physical connection to the PAD may be either by a dial-up connection or by a leased line. After the connection is established the PAD must be informed at what speed you are operating and what initial set of operating parameters you require. On the British Telecom PSS service this is indicated by typing Carriage return twice to indicate the speed, followed by the profile required, followed by another Carriage return, e.g. CR CR V5 CR. The PAD will then respond with identification information and the port number on the PAD and a format effector (FE) which depends on the profile selected, typically Carriage return and Line feed, e.g. GLA/A01 − 231 CRLF.

After the identification from the PAD has been received the user then has two minutes in which to make his NUI known to the PAD. This is achieved by entering 'N' followed by the ten-character NUI and is terminated by carriage return. The network address of the required destination is then entered by typing 'A' followed by the X.25 address of the host computer which the terminal is to be connected with, minus the three-digit country code if the call is within the same network as the PAD. For example, to make a call to the X.25 address 23424126010604 which is in the UK from a PAD, also in the UK, connected to the PSS network, the user would enter 'A24126010604'. Note that the address here can be between 9 and 11 characters in length since the X.25 address may include a subaddress, as is the case in this example. The PAD will reply with the message 'CONN' if the connection request is successful. If any errors occur in the validation of the NIU or the call to the required host fails the error message 'ERR' may be produced. With more user-friendly PAD, such as those sold by some manufacturers of communications equipment, the same connection might be achieved by the following conversation:

PAD>LOGON ABC01	identifying the user to the PAD
PASSWORD?	echo suppressed for security
PAD>CALL STRATHCLYDE	make connection request using mnemonic
CALL CONNECTED . . .	indication that connection was successful

The underlined text is that produced by the PAD, the remainder is entered by the user.

A PAD will generally support more than one terminal connection, in commercially available units 8, 16, or 32 terminals per PAD are typical. The PAD might be based on a microprocessor such as the Z80 or 68000 which would provide sufficient computing power to drive this number of terminals at speeds up to 9600 bps together with an X.25 link at the same speed. Each terminal when connected to a host computer makes use of a separate virtual circuit even if there are other conversations to the same destination, no multiplexing takes place. It is therefore necessary to arrange to have sufficient virtual circuits configured by the supplier of the X.25 network to which the PAD is to be connected upon which outgoing calls may be made to allow for all the terminals to be active.

6.3 SUMMARY OF FACILITIES PROVIDED BY X.3

For each terminal connected to the PAD there are a set of 18 parameters which control the behaviour of the PAD in response to data flowing between the terminal and the PAD. By the alteration of these parameters the host computer to which the user is connected can automatically customise the protocol presented both to the user and to itself. As an alternative the user may himself be able to change the parameter settings to optimise his use of the PAD, or alternatively the PAD may automatically set up parameters whenever a particular terminal is activated or a call made on a particular host. At any time when a host is connected to a terminal session of a PAD it may read the current parameter settings, or set them. A single set-and-read command is defined which is the recommended method of altering PAD parameters from the host, the PAD is then forced to return the values of those parameters which have been set so that the host can validate that the PAD has been able to obey the command given to it.

The PAD parameters are given decimal values from 1 to 18; the actual value is always a single octet. The value of a parameter is generally displayed thus: 1:0 — i.e. parameter 1 has the value 0. We shall now proceed through the various PAD parameters describing their function and the possible values which they can take, together with any interactions with other parameters.

Parameter 1: Escape to PAD command state on receipt of DLE
The default action would be to set this paramter to 1 to indicate that escape to the PAD command state is possible. This would be

used for example if the user wished to terminate a call from the PAD rather than from the host. After sending DLE (typically Control & P) the PAD would respond with a prompt allowing the user to enter a new command such as CLEAR or RESET. If the parameter value is 0 then no escape is possible; this would be useful where transparent input is required from a graphics terminal for example. This parameter is loosely related to parameter 7, which controls the PAD action on receiving a Break signal causing the PAD to enter the command state. If it is not 8 and parameter 1 is 0 then the only way to access the PAD command state is to disconnect the terminal from the PAD and re-access it.

Parameter 2: Character echo

This parameter controls whether the PAD should echo each character back to the terminal as it is typed. The terminal would normally be connected to the PAD in full-duplex mode so that each character is not displayed on the terminal until the PAD has echoed it, thus confirming that the character has been correctly received. This parameter would normally be set to the value of 1 which indicates that echoing is to be performed, it would typically be set to no-echo (i.e. 0) when a password or similar security control was to be entered by the user to gain access to some controlled resource, such as a host computer system or a particular file on such a system.

Parameter 3: Data forwarding characters

When a PAD is accumulating characters sent by a terminal to be sent as a packet to the host, some agreement must be made as to which character or characters will cause the data already entered to be forwarded to the host system which is communicating with the terminal. If the PAD has insufficient buffer space to store further characters received from the terminal and the terminal flow control option is not enabled (see parameter 5) then characters which are being discarded will cause the BEL character to be echoed. Irrespectie of which characters are chosen as data forwarding characters the reception of the 129th character in a sequence of non-forwarding characters will cause the previous 128 characters to be forwarded in a full packet with the More bit set ON to show that further data will follow to complete the message. Once data has been forwarded to the host system, no further editing may take place of that data in the PAD — such as character or buffer delete. Data forwarding may also be under the control of a timer (parameter 4); this will cause the forwarding of any data in the PADs buffer after the timer has expired. The timer is reset each time any character appears in a

previously empty buffer. This is useful in situation where there is
no character which can be reserved as a forwarding character, for
example during graphical input from a graphics terminal each co-
ordinate selected by a light-pen should be forwarded as it is received
as a binary value. Since the coordinates may have any value which
can be stored in an octet it is not possible to predetermine any
forwarding characters to be used in such circumstances. The encoding
of the ranges of characters used for forwarding selected by differing
parameter values are now given:

0	– The 129th character or time-out only
1	– A–Z a–z 0–9
2	– CR
4	– ESC, BEL, ENQ, ACK
8	– DEL, CAN, DC2
16	– ETX, EOT
32	– HT, LF, VT, FF
64	– All other characters less than decimal 32 in value and DEL.

In order to customise the PAD characteristics, the above values of
parameter 3 may be summed to give the desired forwarding characters;
for example, 18 will forward on receipt of CR, ETX or EOT. The
character which caused the data to be forwarded will be sent at the
end of the data which is being forwarded.

Parameter 4: Data forwarding on time-out

As was mentioned above the data forwarding time-out parameter
can allow us to forward data after some period of time has elapsed
with no other forwarding character having arrived. The disadvantage
of this method of data forwarding is the potential high cost to
the user and the high load on the network caused by its use. The
parameter may be 0 to indicate that no time-out will ever occur,
or between 1 and 255 to measure in units of 0.05 seconds the period
of the time-out. For an increase in efficiency it is obviously desirable
to use the highest value possible, though this will reduce the respon-
siveness of interactive software. A typical value might be 10 which
will cause the PAD to buffer data for up to half a second.

Parameter 5: Flow control of the terminal by the PAD.

Where the PAD is connected to an intelligent terminal of some kind
equipped with some data storage medium, paper tape or floppy discs
typically, the rate at which data is sent to the PAD may exceed the
rate at which the date can be forwarded to the host system. This is

fairly unlikely to occur with a human typist unless the network is very congested or the host overloaded. If this parameter is set to 0 then the PAD will discard data which is has no buffer space to contain it and echo a BEL character to the user to warn him that data has been discarded by the PAD. If the parameter is set to 1 then the PAD will send a DC3 (X-OFF) character to the terminal if its buffer is close to being full; when the congestion of buffer space has been alleviated the PAD will send a DC1 (X-ON) to cause the terminal to recommence sending data to the PAD.

Parameter 6: Suppression of service signals from the PAD

PAD service signals are produced in response to unpredictable external events such as the X.25 call being either Cleared or Reset. If the parameter is set to the value 0 then messages related to such events are not to be sent to the terminal; if 1 the terminal will be notified.

Parameter 7: Action on receipt of a Break from the terminal

The Break signal in the title of this parameter is that sequence defined for the particular PAD which is used to signal the host to which the terminal is connected that attention is required. This might be used, for example, to interrupt a long-running program which the user suspects of being faulty, or for halting a listing on the terminal of some file after enough data has been seen. On a standard PAD such as that offered by PSS the Break signal is obtained by Break being signalled by the terminal, i.e. the link is held in the space condition for more than 100 ms. The action of the PAD depends on the setting of this parameter as follows:

0	No action on receipt of Break
1	The PAD transmits an X.25 Interrupt packet to the host
2	The PAD transmits an X.25 Reset packet to the host
5	The PAD transmits an X.25 Interrupt packet and an Indication of Break
8	The PAD leaves the Data Transfer state and waits for a command
21	The action is as in 5 but also sets parameter 8 to 1.

These values allow for varying degrees of sophistication in the handling of Breaks by a host system. Values 1 and 2 simply inform the host quickly than an attention is required, the latter also causing the network to throw away any data in transit between the host and the terminal. In both cases the host would have to reply with the appropriate X.25 response packet. Values of 5 and 21 are less widespread in their destruction of data in transit: the Interrupt packet will be expedited to the host which will then discard all data up to the receipt of the Indication of Break message which will arrive in the normal sequence of packets. The host will reply with an Indication of Break; in the case where value 21 is set the PAD will be discarding all data until it receives this reply, since parameter 8 (which switches off terminal output) is set to 1. When the PAD receives the response Indication of Break it will set parameter 8 back to 0 to cause the PAD to resume sending data to the terminal. This fairly complex mechanism is superior to the simpler methods such as are obtained by values 1 and 2 since, if the terminal is relatively slow, there may be a significant amount of data buffered in the PAD which the terminal will still be forced to receive before the attention requested by the user becomes effective.

Parameter 8: Data delivery to the terminal

When this parameter is set to the value 0 data is delivered to the terminal as fast as it can receive it. When set to 1 all data received by the PAD destined for the terminal is discarded. Receipt of a Reset indication or confirmation causes this parameter to be set to 0.

Parameter 9: Padding after Carriage Return

This parameter controls the number of padding characters which are inserted into the data system by the PAD when a Carriage Return is to be sent to the terminal, either from the host or as an echo to a Carriage Return entered by the user. The padding characters are used to allow simple mechanical terminals whose mechanism requires greater than one character's duration to return the type head across the page before a new line can be printed. If the value is 0 then no padding characters are generated unless the Carriage Return was generated by the PAD under control of parameter 10, in which case either 2 or 4 is generated, depending on whether the terminal is 110 bps or faster. If the parameter is between 1 and 7 the $1-7$ padding characters are generated after a Carriage Return, irrespective of how it was generated.

Parameter 10: Line folding

In order to avoid loss of data if the terminal is less wide across the screen than the host was expecting, the PAD will keep a count of how many non-format-effectors it has sent to the terminal. If this count ever reaches the value of the line folding parameter then the PAD automatically sends a Carriage Return (and possibly some padding characters, depending on parameter 9) to the terminal. The parameter may be either 0 to indicate that the PAD must not insert any format effectors, or between 1 and 255 to indicate when the PAD will wrap the data onto the next line.

Parameter 11: Terminal speed
The CCITT lays down many values for possible terminal speeds, not in any very logical order, some examples are:

    0   110 bps
    2   300 bps
    3   1200 bps

This parameter is 'read-only', the host may enquire of, but not set, this parameter. The value is also sent as an option on the Call Request packet when the virtual circuit is being set up.

Parameter 12: Flow control of the PAD by the terminal
Where the terminal contains some intelligence and perhaps has some auxiliary media such as paper tape punching, the PAD may be told by the terminal to cease sending data in order to allow its peripheral device to catch up, this is effected by transmitting an X-OFF to the PAD, transmission is resumed when the terminal sends an X-ON. This facility is switched on by setting this parameter to 1; if 0 the facility is off. A return to the waiting for command state will also clear a PAD held condition caused by a previous X-OFF.

Parameter 13: Line Feed insertion after Carriage Return
In order to allow for varieties of terminal which automatically insert Line Feeds after Carriage Returns, and the variety of hosts, some of which will insert a Line Feed after a Carriage Return and others not, several values of this parameter are provided so that the most natural effect is obtained on the user's terminal.

    0   No Line Feed insertion is performed
    1   Line Feed is inserted after every Carriage Return transmitted
    4   Line Feed is inserted after every local Carriage Return is echoed
    5   Effect of values 1 plus 4
    6   Line Feed inserted after every Carriage Return echoed to the
         terminal and sent to the host
    7   Effect of values 1 plus 6

Parameter 14: Padding characters inserted after Line Feed
This parameter operates exactly as parameter 9 except that the

effect is to cause the PAD to send padding characters whenever a Line Feed is sent to the terminal if its value is non-zero.

Parameter 15: Editing

This parameter enables or disables the editing of characters awaiting transmission and still held in a buffer in the PAD. If the parameter is set to zero then no editing is possible; if set to 1 then the editing characters described by parameters 16, 17, and 18 apply.

Parameter 16: Character Delete character

If the value of this parameter is 0 then no character deletion of characters stored in the PAD buffer prior to transmission in a packet to the host is possible. If the value is non-zero then the value represents the code of the character to be used to cause the last character in the terminal buffer to be deleted. For example, 8 will cause the ASCII BackSpace character to be used. Parameter 15 must be set to 1 before this parameter is effective.

Parameter 17: Buffer Delete character

As with parameter 16, when non-zero the value of the parameter is the code representing the character which the user may type in order to cause the PAD to discard the entire buffer as generated by the user so far. A typical value would be 24 for the ASCII Cancel character (usually obtained by typing Control/X).

Parameter 18: Buffer Display character

In order that the user can be sure of what characters are in the PADs buffer (for example after repeated use of Character Delete) the Buffer Display character may be entered causing the PAD to display on a new line the current contents of its buffer. As with parameters 16 and 17, parameter 15 must be set to 1 for this character to be operable. If the parameter is zero then no buffer display is possible; if non-zero then the parameter signfies the code to be used. Typically DC2 is used, generally obtained by control/R on a terminal.

6.4 X.29 MAPPING OF PAD COMMANDS AND DATA ONTO X.25

If we assume that the virtual circuit between the PAD and the host to which the terminal wishes to connect is already set up, we can define the way in which ordinary data between the terminal and host is exchanged in X.25 Data packets. As each character is typed on the terminal to be sent to the host it is accumulated in a buffer. If the buffer ever reaches 128 characters then the data is sent as it stands in an X.25 Data packet with the More bit set ON. When one of the characters which has been agreed to be a terminating character is reached, the buffer is sent as an X.25 Data packet with the More bit set OFF. Replies sent by the host are dealt with similarly, except that generally the data is not accumulated one character at a time, but rather each record is sent by an operating system call to the X.25 driver software to be sent to the terminal. If the data will span more than one data packet (128 characters), then the first and any intermediate ones will have the More bit set ON; the final packet will have this bit set OFF.

When a call is to be made from a PAD to a host under instructions from one of the terminals connected with it, the address part of an X.25 Call Request packet is filled in with information obtained from the terminal user, either directly or in the form of a mnemonic which undergoes a translation. The Call User Data field which is at the end of the Call Request and is either 16 or 128 octets in length contains at its start a field termed the Protocol Identifier of four octets. This field is used by X.25 software to decide which high level protocol handler to pass Call Requests to when they arrive. The standard protocol identifier for X.29 is an octet containing one in the first octet. The second octet may contain the PAD profile of the terminal; this is an agreed set of parameter values expressed as an integer. This facility is not of great use since generally the first thing that a host system will do when a connection has been established is to set all the parameters in the most appropriate way for communications with that host. The third octet may contain a copy of parameter 11 indicating the speed of the connection between the PAD and the terminal. The last octet is left zero. Of these only the first is mandatory. The Facility field of the X.25 Call Request may be used to carry network level information, such as a request to make use of reverse charging for the call, or for the call to be a part of a closed user group. Fast Select may also be specified here which would allow for up to 128 octets of Call User Data, of which 124 may be used after the Protocol Identifier has been sent. The remaining Call User Data field (whether it be 12 or 124 octets) is occasionally used by hosts to carry logon information, or where the host which has been called is actually a gateway to another network, the address to be contacted in that network may be carried here.

Whenever the PAD or host wish to exchange some information about the session or alter one of the parameters controlling the behaviour of the terminal they send a PAD control message. This is an X.25 Data message with the Qualified data bit set ON. Before this can be sent any data buffered by the PAD will be forwarded, since X.25 protocol rules dictate that a sequence of packets (i.e. either a single packet with the More bit set OFF or several packets with all but the last having the More bit set ON) must be terminated before a switch can be made between Qualified and Unqualified data messages. The value of the first octet of data contains the message code which signifies the meaning of the data that follows; this is a value between 0 and 6.

Coding of PAD control messages:

Code   Message type

0   Parameters Indication
1   Invitation to Clear
2   Set Parameters
3   Indication of Break
4   Read Parameters
5   Error
6   Set and Read Parameters

The encoding of any further data and the function of these PAD control messages will now be described in the order given above.

Parameters Indication

This returns the values of the parameters in the PAD to the host in response to either a valid Read (Code 4) or Set and Read (Code 6) command, it is also used after a Set (Code 2) to return any settings which the PAD considers to be erroneous. If the Parameters Indication is in response to either a Set and Read or Read Parameters command, then only those fields which were either explicitly set or read will be returned. For those Parameter Indication commands which are returned as a result of a Set command which contains a faulty parameter setting, only the erroneous parameter is returned. An error is signalled by setting the top bit of the parameter number field ON (it is normally set OFF) and setting the value of the returned parameter to zero. As we shall see when we describe the Set and Read commands the parameters are sent as pairs of octets, the first indicating the parameter number, the second the value to which the parameter should be set.

Invitation to Clear

When the X.25 Disconnect Request type packet is issued by either end of a virtual circuit, the effect is to discard any data which is in the network buffered at intermediate nodes etc. and Clear the call as early as possible. This destructive call clear-down is avoided by the use of a message which is carried in the ordinary flow of data which indicates upon arrival that the call may be cleared. The Invitation to Clear is the X.29 implementation of this piece of protocol. In a conversation between a host and a terminal the call is usually ended by the terminal requesting a 'Logoff' from the host when all work has been completed. If the 'Logoff' request caused the host to simply drop the virtual circuit then the terminal user might miss the final messages from the host system (accounting messages produced as a result of issuing the 'Logoff' perhaps) since, although they would be sent into the network, the disconnect would cause them to be discarded. Hence the host will transmit an Invitation to Clear message when the last data message has been sent. Since this message is part of the ordinary flow of data it cannot overtake any data in the network. When the PAD receives the message it will display all the characters from its buffer on the terminal and finally when all the characters have been sent to the terminal it will issue an X.25 level Disconnect Request back to the host, so clearing the X.25 virtual circuit.

Set Parameters

This command is sent by a host to set some or all of the PAD parameters. As mentioned before this message contains pairs of octets, the first of which is the binary value of the parameter number, and the second of which is the value to which the PAD should set that parameter. If the PAD considers that the value to which the host is trying to set a parameter is faulty it will return a Parameters Indication message with the faulty parameters contained in it. The same parameter may appear more than once in the list of parameters to be set (which may themselves be in any order), the last one takes preference in this case. The only limit on how many parameters may be set in a single command is the constraint that the message may not span more than one 128 octet Data packet.

Indication of Break

This control message exists in two forms depending on which indication of Break command in response to receiving a Break command from the terminal and with parameter 7 set to 21, the command consists of three octets. These are the command code of 3, followed by a parameter number and value pair. The parameter passed on the Indication of Break command is always parameter 8, which is used to inform the host that the PAD is discarding all data for the terminal until it is informed otherwise. When the host receives the Interrupt packet sent before the Indication of Break (which may have over-taken data *en route* to the host) it will start to discard any data from the PAD until it receives the Indication of Break message. At this point it knows that any data which follows should no longer be discarded since it was entered after the break was actioned by the terminal user. The PAD continues to discard data destined for the terminal until it receives either an X.25 Reset command or an explicit Set or Set and Read command which resets the value of parameter 8 back to zero. If the host wishes to signal to the terminal a break condition it will send an Indication of Break message; but this will only consist of the command code, no parameter values are sent. This is displayed to the user as a PAD defined message to the user. This is not a facility which is generally made use of.

Read Parameters

In order to establish exactly what the state of a PAD connected terminal is at any point in time the host may issue a Read Parameters command for as many of the parameters as it is interested in. This message has the same format as the Set Parameters command, the message code of 4, followed by pairs of references to parameter values and a dummy value of zero for each one. As many of the parameters as desired may be read, and in any order. The PAD will reply to the Read Parameters command by sending a Parameters Indication command to the host.

Error

If the PAD finds the command it has received to be unacceptable it may return to the host an Error message. This message has a message code of 5 collowed by either one or two octets depending on the type of error being reported.

Code in 1st octet	2nd octet	Description of error
0	none	Received PAD message less than 8 bits long.
1	†	Unrecognised message code.
2	†	Message received incompatible with code.
3	†	Message was not integral number of octets.

† Message code of message in error.

Set and Read Parameters

The Set and Read parameters message type is in the same format as the Set Parameters message in every detail except for the different value of the message code. The effect of this message is to cause the PAD to set those parameters which are mentioned in the command to the values given if possible, and to then return an Indication of Parameters command to the host to show the state of those parameters which were referenced. This is the recommended way in which the host should alter parameters in the PAD so that a check can be made by the host that the desired effect has been achieved at the PAD so the terminal will operate in the way in which the user expects. If unexpected values are returned the host may then decide either to continue or quit depending on which value are set.

Bibliography

British Post Office (1980), PSS Technical User Guide.

British Telecom Study Group Three of PSS Users Forum (1980), A Network Independent Transport Service.

British Telecom Study Group Three of PSS Users Forum (1981), Character Terminal Protocols on PSS.

CCITT (1978), Recommendations X.3, X.28, X.29 on Packet Switched Data Transmission Services, ITU, Geneva.

CCITT (1980), Recommendation X.25, Orange Book Tome VIII.2.

Deasington, R. J. (1984), *A Practical Guide to Computer Communications and Networking*, 2nd edition.

ECMA (1982), Standard ECMA-72 Transport Protocol, 2nd edition.

ISO (1976), Data Communication — High Level Data Link Control Procedures — Frame Structure, ISO 3309.

ISO (1978), Data Communication — High Level Data Link Control Procedures — Elements of Procedure, ISO 4335.

ISO (1983), Reference Model of Open Systems Interconnection, ISO/TC97/SC16, DIS 7498.

ISO (1983), Transport Service Definition, ISO DIS 8072.

ISO (1983), Transport Protocol Specification, ISO DIS 8073.

ISO (1983), Network Service Definition, ISO DIS 8348.

Index